Oxford Lecture Series in
Mathematics and its Applications 13

Series editors
John Ball   Dominic Welsh

OXFORD LECTURE SERIES IN
MATHEMATICS AND ITS APPLICATIONS

1. J. C. Baez (ed.): *Knots and quantum gravity*
2. I. Fonseca and W. Gangbo: *Degree theory in analysis and applications*
3. P. L. Lions: *Mathematical topics in fluid mechanics, Vol. 1: Incompressible models*
4. J. E. Beasley (ed.): *Advances in linear and integer programming*
5. L. W. Beineke and R. J. Wilson (eds.): *Graph connections: Relationships between graph theory and other areas of mathematics*
6. I. Anderson: *Combinatorial designs and tournaments*
7. G. David and S. W. Semmes: *Fractured fractals and broken dreams*
8. Oliver Pretzel: *Codes and algebraic curves*
9. M. Karpinski and W. Rytter: *Fast parallel algorithms for graph matching problems*
10. P. L. Lions: *Mathematical topics in fluid mechanics, Vol. 2: Compressible models*
11. W. T. Tutte: *Graph theory as I have known it*
12. Andrea Braides and Anneliese Defranceschi: *Homogenization of multiple integrals*
13. Thierry Cazenave and Alain Haraux: *An introduction to semilinear evolution equations*
14. J. Y. Chemin: *Perfect incompressible fluids*

An Introduction to Semilinear Evolution Equations

Revised edition

Thierry Cazenave

CNRS and University of Paris VI, France

and

Alain Haraux

CNRS and University of Paris VI, France

Translated by

Yvan Martel

University of Cergy-Pontoise, France

CLARENDON PRESS · OXFORD
1998

Oxford University Press, Great Clarendon Street, Oxford OX2 6DP
Oxford New York
Athens Auckland Bangkok Bogota Buenos Aires Calcutta
Cape Town Chennai Dar es Salaam Delhi Florence Hong Kong Istanbul
Karachi Kuala Lumpur Madrid Melbourne Mexico City Mumbai
Nairobi Paris São Paolo Singapore Taipei Tokyo Toronto Warsaw
and associated companies in
Berlin Ibadan

Oxford is a trade mark of Oxford University Press

Published in the United States
by Oxford University Press Inc., New York

Introduction aux problèmes d'évolution semi-linéaires
© Édition Marketing S.A, 1990
First published by Ellipses

Translation © Oxford University Press, 1998

Aidé par le ministère français chargé de la culture

All rights reserved. No part of this publication may be reproduced,
stored in a retrieval system, or transmitted, in any form or by any means,
without the prior permission in writing of Oxford University Press.
Within the UK, exceptions are allowed in respect of any fair dealing for the
purpose of research or private study, or criticism or review, as permitted
under the Copyright, Designs and Patents Act, 1988, or in the case of
reprographic reproduction in accordance with the terms of licences
issued by the Copyright Licensing Agency. Enquiries concerning
reproduction outside those terms and in other countries should be sent to
the Rights Department, Oxford University Press,
at the address above.

This book is sold subject to the condition that it shall not, by way
of trade or otherwise, be lent, re-sold, hired out, or otherwise circulated
without the publisher's prior consent in any form of binding or cover
other than that in which it is published and without a similar condition
including this condition being imposed on the subsequent purchaser.

A catalogue record for this book is available from the British Library

Library of Congress Cataloging in Publication Data
(Data available)

ISBN 0 19 850277 X (Hbk)

Typeset by Yvan Martel

Printed in Great Britain by
Bookcraft (Bath) Ltd,
Midsomer Norton, Avon

Preface

This book is an expanded version of a post-graduate course taught for several years at the Laboratoire d'Analyse Numérique of the Université Pierre et Marie Curie in Paris. The purpose of this course was to give a self-contained presentation of some recent results concerning the fundamental properties of solutions of semilinear evolution partial differential equations, with special emphasis on the asymptotic behaviour of the solutions.

We begin with a brief description of the abstract theory of semilinear evolution equations, in order to provide the reader with a sufficient background. In particular, we recall the basic results of vector integration (Chapter 1) and linear semigroup theory in Banach spaces (Chapters 2 and 3). Chapter 4 concerns the local existence, uniqueness, and regularity of solutions of abstract semilinear problems.

In Nature, many propagation phenomena are described by evolution equations or evolution systems which may include non-linear interaction or self-interaction terms. In Chapters 5, 6, and 7, we apply some general methods to the following three problems.

(1) The heat equation

$$u_t = \Delta u, \tag{0.1}$$

which models the thermal energy transfer in a homogeneous medium, is the simplest example of a diffusion equation. This equation, as well as the self-interaction problem

$$u_t = \Delta u + f(u), \tag{0.2}$$

can be considered on the entire space $\mathbb{R}^N$ or on various domains Ω (bounded or not) of $\mathbb{R}^N$. In the case in which $\Omega \neq \mathbb{R}^N$, we need to specify a boundary condition on $\Gamma = \partial\Omega$. It can be, for example, a homogeneous Dirichlet condition

$$u = 0 \quad \text{on } \Gamma, \tag{0.3}$$

or a homogeneous Neumann condition

$$\frac{\partial u}{\partial \vec{n}} = 0 \quad \text{on } \Gamma. \tag{0.4}$$

Chapter 5 studies in detail the properties of the solutions of (0.2)–(0.3) when Ω is bounded. In this problem, the maximum principle plays an important role. This is the reason for studying equation (0.2) in the space of continuous functions. Vector-valued generalizations of the form

$$\frac{\partial u_i}{\partial t} = c_i \Delta u_i + f_i(u_1, \ldots, u_k), \quad i = 1, \ldots, k, \tag{0.5}$$

called reaction–diffusion systems, often arise in chemistry and biology. One of the main tools in the study of these systems (and in particular of their non-negative solutions) is the maximum principle, which gives *a priori* estimates in $L^\infty(\Omega)^k$ for the trajectories. We thus develop C_0 methods rather than L^2 methods, which are easier but less suitable in this framework.

(2) The wave equation (also called the Klein–Gordon equation)

$$u_{tt} = \Delta u - mu, \tag{0.6}$$

with $m \geq 0$, models the propagation of different kinds of waves (for example light waves) in homogeneous media. Non-linear models of conservative type arise in quantum mechanics, whereas variants of the form

$$u_{tt} = \Delta u - f(u, u_t) \tag{0.7}$$

appear in the study of vibrating systems with or without damping, and with or without forcing terms. Other perturbations of the wave equation arise in electronics (the telegraph equation, semi-conductors, etc.).

The basic method for studying (0.6) with suitable boundary conditions (for example (0.3)) consists of introducing the associated isometry group in the energy space $H^1 \times L^2$. Local existence and uniqueness of solutions is established in this space. However, in general, the solutions are differentiable only in the sense of the larger space $L^2 \times H^{-1}$. These local questions are considered in Chapter 6.

(3) The Schrödinger equation

$$iu_t = \Delta u, \tag{0.8}$$

possesses a combination of the properties described in (1) and (2). Primarily a simplified model for some problems of optics, this equation also arises in quantum field theory, possibly coupled with the Klein–Gordon equation. Various non-linear perturbations of (0.7) have appeared recently in the study of laser beams when the characteristics of the medium depend upon the temperature; for example, focusing phenomena in some solids (where the medium can break down if the temperature reaches a critical point) and contrastingly, defocusing in a gas medium which weakens the transmitted signal according to the distance from the source. A close examination of sharp properties of solutions of the non-linear Schrödinger equation is delicate, since this problem has a mixed or degenerate nature (neither parabolic nor hyperbolic). In Chapter 7, which is devoted to Schrödinger's equation, it becomes clear that even the local theory requires very elaborate techniques.

The choice of these three problems as model examples is somewhat arbitrary. This selection was motivated by the limited experience of the authors, as well as by the desire to present the easiest models (in particular, semilinear models) for a first approach to the theory of evolution equations. We do not address several other equally worthy problems, such as transport equations, vibrating plates, and fundamental equations of fluid mechanics (such as Boltzmann's equation, the Navier–Stokes equation, etc.). Such complicated systems require many specific methods which could not be covered or even approached in a work of this kind.

Chapters 8, 9, and 10 are devoted to some techniques and results concerning the global behaviour of solutions of semilinear evolution problems as the time variable converges to infinity. In Chapter 8, we establish that, for several kinds of evolution equations, the solutions either blow up in finite time in the original space or they are uniformly bounded in this space for all $t \geq 0$. This is the case for the heat equation and the Klein–Gordon equation with attractive nonlinearity, as well as for non-autonomous problems with dissipation. No such alternative is presently known for Schrödinger's equation. Chapter 9 is devoted to some basic notions of the theory of dynamical systems and its application to models (1) and (2) in an open, bounded domain of $\mathbb{R}^N$. We restrict ourselves to the basic properties, and we give an extensive bibliography for the interested reader. In Chapter 10, we study the asymptotic stability of equilibria. We also discuss the connection between stability and positivity in the case of the heat equation.

Finally, in the notes at the end of each chapter there are various bibliographical comments which provide the reader with a larger overview of the theories discussed. Moreover, the limited character of the examples studied is compensated for by a rather detailed bibliography that refers to similar works. We hope that this bibliography will serve our goal of a sufficient yet comprehensible introduction to the available theory of evolution problems. At the time of publication, new results will have made some parts of this book obsolete. However, we think that the methods presented are, and will continue to be for some years, an indispensable basis for anyone wanting a global view of evolution problems.

Paris   T. C.
1998   A. H.

Contents

 Notation xiii

1. Preliminary results 1

 1.1. Some abstract tools 1
 1.2. The exponential of a linear continuous operator 1
 1.3. Sobolev spaces 2
 1.4. Vector-valued functions 4
  1.4.1. Measurable functions 4
  1.4.2. Integrable functions 7
  1.4.3. The spaces $L^p(I, X)$ 8
  1.4.4. Vector-valued distributions 10
  1.4.5. The spaces $W^{1,p}(I, X)$ 13

2. m-dissipative operators 18

 2.1. Unbounded operators in Banach spaces 18
 2.2. Definition and main properties of m-dissipative operators . . 19
 2.3. Extrapolation 21
 2.4. Unbounded operators in Hilbert spaces 22
 2.5. Complex Hilbert spaces 25
 2.6. Examples in the theory of partial differential equations . . . 26
  2.6.1. The Laplacian in an open subset of $\mathbb{R}^N$: L^2 theory . . . 26
  2.6.2. The Laplacian in an open subset of $\mathbb{R}^N$: C_0 theory . . . 27
  2.6.3. The wave operator (or the Klein–Gordon operator) in $H_0^1(\Omega) \times L^2(\Omega)$ 29
  2.6.4. The wave operator (or the Klein–Gordon operator) in $L^2(\Omega) \times H^{-1}(\Omega)$ 30
  2.6.5. The Schrödinger operator 31

3. The Hille–Yosida–Phillips Theorem and applications 33

 3.1. The semigroup generated by an m-dissipative operator 33
 3.2. Two important special cases 35
 3.3. Extrapolation and weak solutions 38
 3.4. Contraction semigroups and their generators 39
 3.5. Examples in the theory of partial differential equations . . . 42

	3.5.1. The heat equation	42
	3.5.2. The wave equation (or the Klein–Gordon equation)	47
	3.5.3. The Schrödinger equation	47
	3.5.4. The Schrödinger equation in $\mathbb{R}^N$	48

4. Inhomogeneous equations and abstract semilinear problems ... 50

    4.1. Inhomogeneous equations ... 50
    4.2. Gronwall's lemma ... 54
    4.3. Semilinear problems ... 55
        4.3.1. A result of local existence ... 56
        4.3.2. Continuous dependence on initial data ... 59
        4.3.3. Regularity ... 60
    4.4. Isometry groups ... 61

5. The heat equation ... 62

    5.1. Preliminaries ... 62
    5.2. Local existence ... 64
    5.3. Global existence ... 65
    5.4. Blow-up in finite time ... 72
    5.5. Application to a model case ... 76

6. The Klein–Gordon equation ... 78

    6.1. Preliminaries ... 78
        6.1.1. An abstract result ... 78
        6.1.2. Functionals on $H_0^1(\Omega)$ ... 79
    6.2. Local existence ... 82
    6.3. Global existence ... 84
    6.4. Blow-up in finite time ... 87
    6.5. Application to a model case ... 89

7. The Schrödinger equation ... 91

    7.1. Preliminaries ... 91
    7.2. A general result ... 92
    7.3. The linear Schrödinger equation in $\mathbb{R}^N$ ... 95
    7.4. The non-linear Schrödinger equation in $\mathbb{R}^N$: local existence ... 100
        7.4.1. Some estimates ... 101
        7.4.2. Proof of Theorem 7.4.1 ... 106
    7.5. The non-linear Schrödinger equation in $\mathbb{R}^N$: global existence ... 112

7.6. The non-linear Schrödinger equation in $\mathbb{R}^N$: blow up in finite time 114
7.7. A remark concerning behaviour at infinity 120
7.8. Application to a model case 121

8. Bounds on global solutions 124

8.1. The heat equation 124
    8.1.1. A singular Gronwall's lemma: application to the heat equation 125
    8.1.2. Uniform estimates 129
8.2. The Klein–Gordon equation 130
8.3. The non-autonomous heat equation 134
    8.3.1. The Cauchy problem for the non-autonomous heat equation 134
    8.3.2. *A priori* estimates 135
8.4. The dissipative non-autonomous Klein–Gordon equation . . . 137

9. The invariance principle and some applications 142

9.1. Abstract dynamical systems 142
9.2. Liapunov functions and the invariance principle 143
9.3. A dynamical system associated with a semilinear evolution equation 145
9.4. Application to the non-linear heat equation 146
9.5. Application to a dissipative Klein–Gordon equation 149

10. Stability of stationary solutions 154

10.1. Definitions and simple examples 154
10.2. A simple general result 156
10.3. Exponentially stable systems governed by PDE 158
10.4. Stability and positivity 164
    10.4.1. The one-dimensional case 165
    10.4.2. The multidimensional case 167

Bibliography 169

Index 185

Notation

$\mathcal{L}(X,Y)$	the space of linear, continuous mappings from X to Y
$\mathcal{L}(X)$	the space of linear, continuous mappings from X to X
$X^\star$	the topological dual of the vector space X
$D(A)$	the Banach space $(D(A), \|\ \|_{D(A)})$ with $\|u\|_{D(A)} = \|u\| + \|Au\|$, when A is a linear operator with a closed graph
$\mathcal{D}(\Omega)$	the space of C^∞ (real-valued or complex valued) functions with compact support in Ω
$C_0^\infty(\Omega)$	$= C_c^\infty(\Omega) = \mathcal{D}(\Omega)$
$C_c(\Omega)$	the space of continuous functions with compact support in Ω
$C_0(\Omega)$	the space of functions of $C(\overline{\Omega})$ which are zero on $\partial\Omega$
$\mathcal{D}'(\Omega)$	the space of distributions on Ω
$L^p(\Omega)$	the space of measurable functions on Ω such that $\|u\|^p$ is integrable $(1 \leq p < \infty)$
$\|u\|_{L^p}$	$= \left(\int_\Omega |u|^p\right)^{1/p}$, for $u \in L^p(\Omega)$
$L^\infty(\Omega)$	the space of measurable functions u on Ω such that there exists C such that $|u(x)| \leq C$ for almost every $x \in \Omega$
$\|u\|_{L^\infty}$	$= \mathrm{Inf}\{C > 0, |u(x)| \leq C \text{ almost everywhere}\}$, for $u \in L^\infty(\Omega)$
p'	the conjugate exponent of p, i.e. $p' = p/(p-1)$ for $1 \leq p \leq \infty$
D^α	$= \dfrac{\partial^{|\alpha|}}{\partial^{\alpha_1} x_1 \ldots \partial^{\alpha_N} x_N}$, $\alpha = (\alpha_1, \ldots, \alpha_N)$, $|\alpha| = \sum\limits_{i=1}^N \alpha_i$
$W^{m,p}(\Omega)$	$= \{f \in L^p(\Omega), D^\alpha f \in L^p(\Omega) \text{ for all } \alpha \in \mathbb{N}^N \text{ such that } |\alpha| \leq m\}$
$\|u\|_{W^{m,p}}$	$= \sum_{|\alpha| \leq m} \|D^\alpha u\|_{L^p}$ for $u \in W^{m,p}(\Omega)$
$W_0^{m,p}(\Omega)$	the closure of $\mathcal{D}(\Omega)$ with respect to the norm $\|\ \|_{W^{m,p}}$
$H^m(\Omega)$	$= W^{m,2}(\Omega)$
$\|u\|_{H^m}$	$= \left(\sum_{|\alpha| \leq m} (\|D^\alpha u\|_{L^2})^2\right)^{1/2}$ for $u \in H^m(\Omega)$

Notation

$H_0^m(\Omega)$ $= W_0^{m,2}(\Omega)$

$\mathcal{D}(I, X)$ the space of C^∞ functions with compact support from I to X

u' $= u_t = du/dt$, for $u \in \mathcal{D}'(I, X)$

$C_c(I, X)$ the space of continuous functions with compact support from I to X

$C_b(I, X)$ the space of continuous and bounded functions from I to X

$C_{b,u}(I, X)$ the space of uniformly continuous and bounded functions from I to X

$L^p(I, X)$ the space of measurable functions u on I with values in X and such that $\|u\|^p$ is integrable ($1 \leq p < \infty$)

$\|u\|_{L^p}$ $= \left(\int_I |u|^p\right)^{1/p}$, for $u \in L^p(I, X)$

$L^\infty(I, X)$ the space of measurable functions u on I such that there exists C such that $\|u(x)\| \leq C$ for almost every $x \in I$

$\|u\|_{L^\infty}$ $= \text{Inf}\{C > 0, |u(x)| \leq C \text{ almost everywhere}\}$, for $u \in L^\infty(I, X)$

$W^{1,p}(I, X) = \{u \in L^p(I, X),\ u' \in L^p(I, X), \text{ in the sense of } \mathcal{D}'(I, X)\}$

$\|u\|_{W^{1,p}}$ $= \|u\|_{L^p} + \|u'\|_{L^p}$ for $u \in W^{1,p}(I, X)$

1
Preliminary results

1.1. Some abstract tools

We recall here some classical theorems of functional analysis that are necessary for the study of semilinear evolution equations. The proofs can be found in Brezis [2].

Theorem 1.1.1. (The Banach Fixed Point Theorem) *Let (E, d) be a complete metric space and let $f : E \to E$ be a mapping such that there exists $k \in [0, 1)$ satisfying $d(f(x), f(y)) \le k d(x, y)$ for all $(x, y) \in E \times E$. Then there exists a unique point $x_0 \in E$ such that $f(x_0) = x_0$.*

Theorem 1.1.2. (The Closed Graph Theorem) *Let X and Y be Banach spaces and let $A : X \to Y$ be a linear mapping. Then $A \in \mathcal{L}(X, Y)$ if and only if the graph of A is a closed subspace of $X \times Y$.*

Remark 1.1.3. We recall that the graph of A is $G(A) = \{(x, y) \in X \times Y;\ y = Ax\}$.

Theorem 1.1.4. (The Lax–Milgram Theorem) *Let H be a Hilbert space and let $a : H \times H \to \mathbb{R}$ be a bilinear functional. Assume that there exist two constants $C < \infty$, $\alpha > 0$ such that:*

(i) $|a(u, v)| \le C \|u\| \|v\|$ for all $(u, v) \in H \times H$ (continuity);

(ii) $a(u, u) \ge \alpha \|u\|^2$ for all $u \in H$ (coerciveness).

Then, for every $f \in H^*$ (the dual space of H), there exists a unique $u \in H$ such that $a(u, v) = \langle f, v \rangle$ for all $v \in H$.

1.2. The exponential of a linear continuous operator

Let X be a Banach space and let $A \in \mathcal{L}(X)$.

Definition 1.2.1. We denote by e^A the sum of the series $\sum_{n \ge 0} \frac{1}{n!} A^n$.

It is clear that the series is norm convergent in $\mathcal{L}(X)$ and that $\|e^A\| \le e^{\|A\|}$. Furthermore, it is well known that if A and B commute, then $e^{A+B} = e^A e^B$.

2 Preliminary results

In addition, for fixed A, the function $t \mapsto e^{tA}$ belongs to $C^\infty(\mathbb{R}, \mathcal{L}(X))$ and we have
$$\frac{\mathrm{d}}{\mathrm{d}t}e^{tA} = e^{tA}A = Ae^{tA},$$
for all $t \in \mathbb{R}$. Finally, we have the following classical result.

Proposition 1.2.2. *Let $A \in \mathcal{L}(X)$. For all $T > 0$ and all $x \in X$, there exists a unique solution $u \in C^1([0,T], X)$ of the following problem:*
$$\begin{cases} u'(t) = Au(t), & \text{for all } t \in [0,T]; \\ u(0) = x. \end{cases}$$
This solution is given by $u(t) = e^{tA}x$, for all $t \in [0,T]$.

Proof. It is clear that $e^{tA}x$ is a solution therefore, we need only show uniqueness. Let v be another solution and let $z(t) = e^{-tA}v(t)$. We have
$$z'(t) = e^{-tA}(Av(t)) - A(e^{-tA}v(t)) = 0.$$
Therefore, $z(t) \equiv z(0) = x$; and so $v(t) = e^{tA}x$. □

1.3. Sobolev spaces

We refer to Adams [1] for the proofs of the results given below. Consider an open subset Ω of $\mathbb{R}^N$. A distribution $T \in \mathcal{D}'(\Omega)$ is said to belong to $L^p(\Omega)$ ($1 \leq p \leq \infty$) if there exists a function $f \in L^p(\Omega)$ such that
$$\langle T, \varphi \rangle = \int_\Omega f(x)\varphi(x)\,\mathrm{d}x,$$
for all $\varphi \in \mathcal{D}(\Omega)$. In that case, it is well known that f is unique. Let $m \in \mathbb{N}$ and let $p \in [1, \infty]$. Define
$$W^{m,p}(\Omega) = \{f \in L^p(\Omega), D^\alpha f \in L^p(\Omega) \text{ for all } \alpha \in \mathbb{N}^m \text{ such that } |\alpha| \leq m\}.$$
$W^{m,p}(\Omega)$ is a Banach space when equipped with the norm defined by
$$\|f\|_{W^{m,p}} = \sum_{|\alpha| \leq m} \|D^\alpha f\|_{L^p},$$
for all $f \in W^{m,p}(\Omega)$. For all m, p as above, we denote by $W_0^{m,p}(\Omega)$ the closure of $\mathcal{D}(\Omega)$ in $W^{m,p}(\Omega)$. If $p = 2$, one sets $W^{m,2}(\Omega) = H^m(\Omega)$, $W_0^{m,2}(\Omega) = H_0^m(\Omega)$ and one equips $H^m(\Omega)$ with the following equivalent norm:
$$\|f\|_{H^m} = \left(\sum_{|\alpha| \leq m} \|D^\alpha u\|_{L^2}^2 \right)^{\frac{1}{2}}.$$

Then $H^m(\Omega)$ is a Hilbert space with the scalar product
$$\langle u, v \rangle_{H^m} = \sum_{|\alpha| \le m} \int_\Omega D^\alpha u D^\alpha v \, dx.$$
If Ω is bounded, there exists a constant $C(\Omega)$ such that
$$\|u\|_{L^2} \le C(\Omega) \|\nabla u\|_{L^2},$$
for all $u \in H_0^1(\Omega)$ (this is Poincaré's inequality). It may be more convenient to equip $H_0^1(\Omega)$ with the following scalar product
$$\langle u, v \rangle = \int_\Omega \nabla u \cdot \nabla v \, dx,$$
which defines an equivalent norm to $\|\cdot\|_{H^1}$ on the closed space $H_0^1(\Omega)$. The following two results are essential in the theory of partial differential equations.

Theorem 1.3.1. *If Ω is open and has a Lipschitz continuous boundary, then:*

(i) *if $1 \le p < N$, then $W^{1,p}(\Omega) \hookrightarrow L^q(\Omega)$, for every $q \in [p, p^*]$, where $p^* = Np/(N-p)$;*

(ii) *if $p = N$, then $W^{1,p}(\Omega) \hookrightarrow L^q(\Omega)$, for every $q \in [p, \infty)$;*

(iii) *if $p > N$, then $W^{1,p}(\Omega) \hookrightarrow L^\infty(\Omega) \cap C^{0,\alpha}(\Omega)$, where $\alpha = (p-N)/p$.*

Theorem 1.3.2. *In addition, if Ω is bounded, embeddings (ii) and (iii) of Theorem 1.3.1 are compact. Embedding (i) is compact for $q \in [p, p^*)$.*

Remark 1.3.3. The conclusions of Theorems 1.3.1 and 1.3.2 remain valid without any smoothness assumption on Ω, if one replaces $W^{1,p}(\Omega)$ by $W_0^{1,p}(\Omega)$.

We also recall the following result (see Friedman [1], Theorem 9.3, p. 24).

Theorem 1.3.4. *Let q, r be such that $1 \le q, r \le \infty$, and let j, m be integers, $0 \le j < m$. Let $a \in [j/m, 1]$ ($a < 1$ if $m - j - N/r$ is an integer ≥ 0), and let p be given by*
$$\frac{1}{p} = \frac{j}{N} + a\left(\frac{1}{r} - \frac{m}{N}\right) + (1-a)\frac{1}{q}.$$
Then there exists $C(q, r, j, m, a, n)$ such that
$$\sum_{|\alpha|=j} \|D^\alpha u\|_{L^p} \le C \left(\sum_{|\alpha|=m} \|D^\alpha u\|_{L^r} \right)^a \|u\|_{L^q}^{1-a},$$
for every $u \in \mathcal{D}(\mathbb{R}^N)$.

Finally, we recall the following composition rule (see Marcus and Mizel [1]).

Proposition 1.3.5. *Let $F : \mathbb{R} \to \mathbb{R}$ be a Lipschitz continuous function, and let $1 \leq p \leq \infty$. Then, for all $u \in W^{1,p}(\Omega)$, the function $F(u)$ belongs to $W^{1,p}\Omega)$. Moreover, if N is the set of points where F is not differentiable ($|N| = 0$), then*

$$\nabla F(u) = \begin{cases} F'(u) \nabla u, & \text{if } u \notin N; \\ 0, & \text{if } u \in N; \end{cases}$$

almost everywhere in Ω.

In particular, we have the following result.

Corollary 1.3.6. *Let $1 \leq p \leq \infty$. For all $u \in W^{1,p}(\Omega)$, we have $u^+, u^-, |u| \in W^{1,p}(\Omega)$. Moreover,*

$$\nabla(u^+) = \begin{cases} \nabla u, & \text{if } u > 0; \\ 0, & \text{if } u \leq 0; \end{cases}$$

almost everywhere.

1.4. Vector-valued functions

We present here some results on vector integration and vector-valued distributions that will be useful throughout this book. We consider a Banach space X and an open interval $I \subset \mathbb{R}$.

1.4.1. Measurable functions

Definition 1.4.1. A function $f : I \to X$ is measurable if there exists a set $E \subset I$ of measure 0 and a sequence $(f_n)_{n \geq 0} \subset C_c(I, X)$ such that $f_n(t) \to f(t)$ as $n \to \infty$, for all $t \in I \setminus E$.

Remark 1.4.2. If $f : I \to X$ is measurable, then $\|f\| : I \to \mathbb{R}$ is measurable.

Proposition 1.4.3. *Let $(f_n)_{n \geq 0}$ be a sequence of measurable functions $I \to X$ and let $f : I \to X$ be such that $f_n(t) \to f(t)$ as $n \to \infty$, for almost all $t \in I$. Then f is measurable.*

Proof. $f_n \to f$ on $I \setminus E$ with $|E| = 0$. Let $(f_{n,k})_{k \geq 0}$ be a sequence of continuous functions with compact supports such that $f_{n,k} \to f_n$ almost everywhere as $k \to \infty$. By applying Egorov's theorem to the sequence $\|f_{n,k} - f_n\|$, we obtain the existence of $E_n \subset I$ with $|E_n| \leq 2^{-n}$, such that $f_{n,k} \to f_n$ uniformly on

$I\setminus E_n$. Let $k(n)$ be such that $\|f_{n,k(n)} - f_n\| \leq 1/n$ on $I\setminus E_n$ and let $g_n = f_{n,k(n)}$. Take $F = E \cup \left(\bigcap_{m\geq 0} \bigcup_{n>m} E_n \right)$ then $|F| = 0$. Let $t \in I\setminus F$. We have $f_n(t) \to f(t)$; on the other hand, for n large enough, $t \in I\setminus E_n$. It follows that $\|g_n - f_n\| \leq 1/n$. Therefore, $g_n(t) \to f(t)$ and so f is measurable. □

Remark 1.4.4. If $f : I \to X$ and $\varphi : I \to \mathbb{R}$ are measurable, then $\varphi f : I \to X$ is measurable.

Remark 1.4.5. If $(x_n)_{n\geq 0}$ is a family of elements of X and if $(\omega_n)_{n\geq 0}$ is a family of measurable subsets of I such that $\omega_i \cap \omega_j = \emptyset$ for $i \neq j$, then $\sum_{n\geq 0} x_n 1_{\omega_n}$ is measurable.

Proposition 1.4.6. (Pettis' Theorem) Consider $f : I \to X$. Then f is measurable if and only if the following two conditions are satisfied:

(i) f is weakly measurable (i.e. for every $x' \in X^\star$, the function $t \mapsto \langle x', f(t) \rangle$ is measurable);

(ii) there exists a set $N \subset I$ of measure 0 such that $f(I \setminus N)$ is separable.

Proof. First, since f is measurable, it is clear that f is weakly measurable. Now let $(f_n)_{n\geq 0} \subset C_c(I, X)$ be a sequence such that $f_n \to f$ on $I\setminus N$ as $n \to \infty$, where $|N| = 0$. It is clear that $f_n(I \setminus N)$ is separable, and then so is $f(I \setminus N)$.

Conversely, we may assume that $f(I)$ is separable, so that X is separable (by possibly replacing X by the smallest closed subspace of X containing $f(I)$). We need the following lemma (see Yosida [1], p. 132).

Lemma 1.4.7. Let X be a separable Banach space, let $X^\star$ be its dual, and let $S^\star$ be the unit ball of $X^\star$. There exists a sequence $(x'_n)_{n\geq 0}$ of $S^\star$ such that, for every $x' \in S^\star$, there exists a subsequence $(x'_{n_k})_{k\geq 0}$ of $(x'_n)_{n\geq 0}$ with $x'_{n_k}(x) \to x'(x)$ for all $x \in X$.

Proof. Let $(x_n)_{n\geq 0}$ be dense in X. For all $n \geq 0$, define $F_n : S^\star \to \ell^2(n)$, by

$$F_n(x') = (x'(x_1), \ldots, x'(x_n)),$$

for all $x' \in X^\star$. Since $\ell^2(n)$ is separable, there exists a sequence $(x'_{n_k})_{k\geq 0}$ of $S^\star$ such that $F_n((x'_{n_k})_{k\geq 0})$ is dense in $F_n(S^\star)$. In particular, for all $x' \in X^\star$, there exists $x'_{n_{k(n)}} \in S^\star$ such that

$$|x'(x_j) - x'_{n_{k(n)}}(x_j)| \leq \frac{1}{n},$$

for $1 \leq j \leq n$. It follows that $x'_{n_{k(n)}}(x_j) \to x'(x_j)$ as $n \to \infty$, for all $j \in \mathbb{N}$. Since $(x_n)_{n \geq 0}$ is dense in X, we deduce easily that $x'_{n_{k(n)}}(x) \to x'(x)$ as $n \to \infty$, for all $x \in X$. The result follows. □

End of the proof of Proposition 1.4.6. Let $x \in X$. Then $t \mapsto \|f(t) - x\|$ is measurable. Indeed, for all $a \geq 0$,
$$\{t, \|f(t) - x\| \leq a\} = \bigcap_{\|x'\| \leq 1} \{t, |x'(f(t) - x)| \leq a\};$$
and it follows from Lemma 1.4.7 that
$$\{t, \|f(t) - x\| \leq a\} = \bigcap_{n \geq 0} \{t, |x_n'(f(t) - x)| \leq a\}.$$
Since the set on the right-hand side is clearly measurable, $t \mapsto \|f(t) - x\|$ is measurable. Now consider $n \geq 0$. The set $f(I)$ (which is separable) can be covered by a countable union of balls $B(x_j)$ of centres x_j and radius $1/n$. Consider $f_n : I \to X$, defined by
$$f_n = \sum x_j 1_{\omega_j},$$
where $\omega_0 = \{t;\, f(t) \in B(x_0)\}$ and $\omega_j = \{t;\, f(t) \in B(x_j)\} \setminus \bigcup_{0 \leq i < j} B(x_i)$. By Remark 1.4.5, f_n is measurable, and it is clear by construction that $\|f_n(t) - f(t)\| \leq 1/n$ for all $t \in I$. Therefore, it follows from Proposition 1.4.3 that f is measurable. □

Corollary 1.4.8. Let $f : I \to X$ be weakly continuous (i.e. if $t_n \to t$, then $f(t_n) \rightharpoonup f(t)$ weakly in X). Then f is measurable.

Proof. By Proposition 1.4.6, it is sufficient to prove that $f(I)$ is separable. Let A be the convex hull of $f(I \cap \mathbb{Q})$ and let $\overline{A}$ be the weak closure of A. We have $f(I) \subset \overline{A}$. On the other hand, $\overline{A}$ is also the strong closure of A and so $\overline{A}$ is separable. □

Corollary 1.4.9. Let $(f_n)_{n \geq 0}$ be a sequence of measurable functions $I \to X$ and let $f : I \to X$. If, for almost every $t \in I$, $f_n(t) \rightharpoonup f(t)$ as $n \to \infty$, then f is measurable.

Proof. Let $x' \in X^*$. Since $\langle x', f_n(t) \rangle \to \langle x', f(t) \rangle$ almost everywhere, the function $t \mapsto \langle x', f(t) \rangle$ is measurable, and so f is weakly measurable.

On the other hand, for every $n \in \mathbb{N}$, there exists a set E_n of measure 0 such that $f_n(I \setminus E_n)$ is separable. Consider the set $E = \bigcup_{n \geq 1} E_n$, $|E| = 0$. Let A be the convex hull of $\bigcup_{n \geq 1} f_n(I \setminus E)$ and let $\overline{A}$ be the weak closure of A. We have $f(I \setminus E) \subset \overline{A}$. Furthermore, $\overline{A}$ is also the strong closure of A and so $\overline{A}$ is separable. It follows that f is measurable. □

1.4.2. Integrable functions

Definition 1.4.10. A measurable function $f : I \to X$ is integrable if there exists a sequence $(f_n)_{n\geq 0} \subset C_c(I, X)$ such that

$$\int_I \|f_n(t) - f(t)\| \, dt \to 0$$

as $n \to \infty$.

Remark 1.4.11. $\|f_n - f\|$ is non-negative and measurable, and so $\int_I \|f_n - f\|$ makes sense.

Proposition 1.4.12. Let $f : I \to X$ be integrable. There exists $x \in X$ such that if a sequence $(f_n)_{n\geq 0} \subset C_c(I, X)$ satisfies $\int_I \|f_n - f\| \longrightarrow 0$, as $n \to \infty$ then one has $\int_I f_n \longrightarrow x$ as $n \to \infty$.

Proof. We have

$$\left\| \int_I f_n - \int_I f_p \right\| \leq \int_I \|f_n - f\| + \int_I \|f_p - f\|.$$

Therefore, $\int_I f_n$ is a Cauchy sequence that converges to some element $x \in X$. Consider another sequence $(g_n)_{n\geq 0}$ that satisfies $\int \|g_n - f\| \longrightarrow 0$ as $n \to \infty$. We have

$$\left\| \int_I g_n - x \right\| \leq \int_I \|g_n - f\| + \int_I \|f_n - f\| + \left\| \int_I f_n - x \right\|.$$

Therefore, $\int_I g_n \longrightarrow x$ as $n \to \infty$. □

Definition 1.4.13 The element x constructed above is denoted by $\int f$, or $\int_I f$. If $I = (a, b)$, it is also denoted by $\int_a^b f$. As for real-valued functions, it is convenient to set $\int_b^a f = -\int_a^b f$.

Proposition 1.4.14. (Bochner's Theorem) Let $f : I \to X$ be measurable. Then f is integrable if and only if $\|f\|$ is integrable. In addition, we have

$$\left\| \int_I f \right\| \leq \int_I \|f\|.$$

Proof. Assume that f is integrable and consider a sequence $(f_n)_{n\geq 0} \subset C_c(I, X)$ such that $\int_I \|f_n - f\| \to 0$. We have $\|f\| \leq \|f_n\| + \|f_n - f\|$; and so $\|f\|$ is integrable.

Conversely, suppose that $\|f\|$ is integrable. Let $g_n \in C_c(I, \mathbb{R})$ be a sequence such that $g_n \to \|f\|$ in $L^1(I)$ and such that $|g_n| \leq g$ almost everywhere, for some $g \in L^1(I)$. Let $(f_n)_{n\geq 0} \subset C_c(I, X)$ be a sequence such that $f_n \to f$ almost everywhere. Finally, let

$$u_n = \frac{|g_n|}{\|f_n\| + \frac{1}{n}} f_n.$$

We have $\|u_n\| \leq g \in L^1(I)$ and $u_n \to f$ almost everywhere. Therefore, we have $\int_I \|u_n - f\| \to 0$ and so f is integrable. Finally, it follows from Fatou's lemma that

$$\left\| \int_I f \right\| \leq \lim_{n \to \infty} \left\| \int_I u_n \right\| \leq \lim_{n \to \infty} \int_I \|u_n\| \leq \int_I \|f\|,$$

which completes the proof. □

Corollary 1.4.15. (The Dominated Convergence Theorem) *Let $(f_n)_{n\geq 0}$ be a sequence of integrable functions $I \to X$, let $g : I \to \mathbb{R}$ be an integrable function, and let $f : I \to X$. If*

$$\begin{cases} \text{for all } n \in \mathbb{N}, \|f_n\| \leq g, & \text{almost everywhere on } I, \\ f_n(t) \longrightarrow f(t) \text{ as } n \to \infty, & \text{for almost all } t \in I, \end{cases}$$

then f is integrable and $\int_I f = \lim_{n \to \infty} \int_I f_n$.

1.4.3. The spaces $L^p(I, X)$

Definition 1.4.16. Let $p \in [1, \infty]$. One denotes by $L^p(I, X)$ the set of (equivalence classes of) measurable functions $f : I \to X$ such that $t \mapsto \|f(t)\|$ belongs to $L^p(I)$. For $f \in L^p(I, X)$, one defines

$$\|f\|_{L^p} = \begin{cases} \left(\int \|f(t)\|^p \, dt \right)^{\frac{1}{p}}, & \text{if } p < \infty; \\ \operatorname{Ess\,sup}_{t \in I} \|f(t)\|, & \text{if } p = \infty. \end{cases}$$

Proposition 1.4.17. *$(L^p(I, X), \|\cdot\|_{L^p})$ is a Banach space. If $p < \infty$, then $\mathcal{D}(I, X)$ is dense in $L^p(I, X)$.*

Proof. The proof is similar to that of the real-valued case (in particular, the density of $\mathcal{D}(I, X)$ is obtained by truncation and convolution). □

Remark 1.4.18. Let $f \in L^p(I, X)$ and let $g \in L^{p'}(I, X^*)$. Then

$$t \mapsto \langle g(t), f(t) \rangle_{X^*, X}$$

is integrable and
$$\int |\langle g(t), f(t)\rangle_{X^*,X}| \leq \|f\|_{L^p}\|g\|_{L^{p'}}.$$

The following result is related to the preceding remark. The proof is much more difficult than that for real-valued functions.

Theorem 1.4.19. If $1 \leq p < \infty$ and if X is reflexive or if X^* is separable, then $(L^p(I,X))^* \approx L^{p'}(I,X^*)$. In addition, if $1 < p < \infty$ and if X is reflexive, then $L^p(I,X)$ is reflexive.

Proof. See Dinculeanu [1] (p. 252, Ch. 13, Cor. 1 of Thm 8). □

Remark 1.4.20. If I is bounded and if $1 \leq q \leq p \leq \infty$, then $L^p(I,X) \hookrightarrow L^q(I,X)$.

Definition 1.4.21. Let $1 \leq p \leq \infty$. We denote by $L^p_{\text{loc}}(I,X)$ the set of measurable functions $f : I \to X$ such that for every compact interval $J \subset I$, $f_{|J} \in L^p(J,X)$.

Proposition 1.4.22. Let X and Y be Banach spaces and let $A \in \mathcal{L}(X,Y)$. If $f \in L^p(I,X)$, then $Af \in L^p(I,Y)$ and $\|Af\|_{L^p} \leq \|A\|_{\mathcal{L}(X,Y)}\|f\|_{L^p}$. If $f \in L^1(I,X)$, then $A\left(\int_I f\right) = \int_I Af$.

Proof. First, assume that $p < \infty$. The result is well known for $f \in \mathcal{D}(I,X)$, and the general case follows from a density argument (Proposition 1.4.17). If $p = \infty$, it is clear that Af is measurable, and that, for almost all $t \in I$,
$$\|Af(t)\| \leq \|A\|_{\mathcal{L}(X,Y)}\|f(t)\| \leq \|A\|_{\mathcal{L}(X,Y)}\|f\|_{L^\infty};$$
hence the result. □

Corollary 1.4.23. If $X \hookrightarrow Y$ and if $f \in L^1(I,X)$, then $f \in L^1(I,Y)$ and the integrals of f in the sense of X and Y coincide.

Proposition 1.4.24. Let $1 \leq p \leq \infty$. Let $(f_n)_{n \geq 0}$ be a bounded sequence of $L^p(I,X)$ and let $f : I \to X$ be such that $f_n(t) \rightharpoonup f(t)$ weakly in X as $n \to \infty$, for almost all $t \in I$. Then $f \in L^p(I,X)$, and $\|f\|_{L^p} \leq \liminf_{n \to \infty} \|f_n\|_{L^p}$.

Proof. By Corollary 1.4.9, f is measurable. We define g_n and g by
$$g_n(t) = \inf_{k \geq n} \|f_k(t)\|;$$
$$g(t) = \lim_{n \to \infty} g_n(t)$$
$$g(t) = \liminf_{n \to \infty} \|f_n(t)\| \quad \text{almost everywhere.}$$

Since $g_n(t) \leq \|f_n(t)\|$ almost everywhere, it follows from the monotone convergence theorem that g is integrable and that $\|g\|_{L^p} = \lim_{n\to\infty} \|g_n\|_{L^p}$. By weak lower semicontinuity of the norm, we have

$$\|f\|_{L^p(I,X)} \leq \|g\|_{L^p} = \lim_{n\to\infty} \|g_n\|_{L^p} \leq \liminf_{n\to\infty} \|f_n\|_{L^p(I,X)};$$

hence the result. □

1.4.4. Vector-valued distributions

Definition 1.4.25. We denote by $\mathcal{D}'((I,X))$ the space $\mathcal{L}(\mathcal{D}(I), X)$. It is called the space of X-valued distributions on I.

Remark 1.4.26. For the definition of $\mathcal{D}(I)$ and its topology, see Schwartz [1].

Remark 1.4.27. Let $f \in L^1_{\mathrm{loc}}(I, X)$. Then

$$\langle T_f, \varphi \rangle = \int_I f\varphi,$$

for $\varphi \in \mathcal{D}(I)$, defines a distribution $T_f \in \mathcal{D}'(I, X)$. We will sometimes denote by f the distribution T_f.

Definition 1.4.28. Let $T \in \mathcal{D}'(I, X)$. We define the distributional derivative of T, $T' \in \mathcal{D}'(I, X)$, by

$$\langle T', \varphi \rangle = -\langle T, \varphi' \rangle,$$

for $\varphi \in \mathcal{D}(I)$.

Proposition 1.4.29. Let $1 \leq p < \infty$ and let $f \in L^p(\mathbb{R}, X)$. Let

$$T_h f(t) = \frac{1}{h} \int_t^{t+h} f(s)\,ds,$$

for $t \in \mathbb{R}$ and $h \neq 0$. Then $T_h f \in L^p(\mathbb{R}, X) \cap C_b(\mathbb{R}, X)$ and $T_h f \to f$ in $L^p(\mathbb{R}, X)$ and almost everywhere as $h \to 0$.

Proof. By the dominated convergence theorem, it is clear that $T_h f \in C(\mathbb{R}, X)$. Furthermore, by Hölder's inequality, we have

$$\|T_h f(t)\|^p \leq \frac{1}{h} \int_t^{t+h} \|f(s)\|^p \, ds.$$

Hence

$$\int_{\mathbb{R}} \|T_h f(t)\|^p \, dt \leq \frac{1}{h} \int_{\mathbb{R}} \int_t^{t+h} \|f(s)\|^p \, ds \, dt$$

$$\leq \frac{1}{h} \int_0^h \int_{\mathbb{R}} \|f(t)\|^p \, dt \, ds \leq \int_{\mathbb{R}} \|f(t)\|^p \, dt.$$

It follows that $T_h \in \mathcal{L}(L^p(\mathbb{R}, X))$ and that $\|T_h\| \leq 1$. Let $A_h = T_h - I$. One has $\|A_h\|_{\mathcal{L}(L^p)} \leq 2$. Let $(f_n)_{n \geq 1} \subset \mathcal{D}(\mathbb{R}, X)$ be a sequence such that $f_n \to f$ as $n \to \infty$ in $L^p(\mathbb{R}, X)$ and on $\mathbb{R} \setminus E$, with $|E| = 0$ (such a sequence exists by Proposition 1.4.17). Let $t \in \mathbb{R} \setminus E$. We have

$$\|A_h f(t)\| \leq \|A_h(f(t) - f_n(t))\| + \|A_h f_n(t)\|$$

$$\leq \|f(t) - f_n(t)\| + \frac{1}{h} \int_t^{t+h} \|f(s) - f_n(s)\| \, ds + \|A_h f_n(t)\|.$$

Let $\varepsilon > 0$. For n large enough, one has $\|f(t) - f_n(t)\| \leq \varepsilon/4$.

On the other hand, since $\|f(\cdot) - f_n(\cdot)\| \in L^1_{\text{loc}}(\mathbb{R})$, by the theory of Lebesgue points (see Dunford and Schwartz [1], p. 217, Theorem 8) we know that

$$\frac{1}{h} \int_t^{t+h} \|f(s) - f_n(s)\| \, ds \longrightarrow \|f(t) - f_n(t)\|,$$

for almost all $t \in \mathbb{R}$, as $h \to 0$. Therefore, for almost all t, n being fixed so that $\|f(t) - f_n(t)\| \leq \varepsilon/4$, and if h is small enough, we have

$$\frac{1}{h} \int_t^{t+h} \|f(s) - f_n(s)\| \, ds \leq \varepsilon/2.$$

Finally, since $f_n \in \mathcal{D}(\mathbb{R}, X)$, for h small enough, we find that

$$\|A_h f_n(t)\|_{L^p} \leq \varepsilon/4.$$

Therefore, for almost all t and for h sufficiently small, we have $\|A_h f(t)\|_{L^p} \leq \varepsilon$. Taking $\varepsilon = 1/n$, we obtain $\|A_h f(t)\|_{L^p} \leq 1/n$ if h is small enough, for all $t \in \mathbb{R} \setminus E_n$, where the measure of E_n is 0. It follows that $T_h f \to f$ as $h \to 0$, for all $t \in \mathbb{R} \setminus \bigcup_{n \geq 1} E_n$, i.e. almost everywhere. Furthermore, we have

$$\|A_h f\|_{L^p(\mathbb{R}, X)} \leq 2\|f - f_n\|_{L^p(\mathbb{R}, X)} + \|A_h f_n\|_{L^p(\mathbb{R}, X)}.$$

Given any n, it is well known that $\|A_h f_n\|_{L^p(\mathbb{R}, X)} \to 0$ as $h \to 0$; it follows that $\|A_h f\|_{L^p(\mathbb{R}, X)} \to 0$, which completes the proof. □

Corollary 1.4.30. Let $f \in L^1_{\text{loc}}(I, X)$ be such that $f = 0$ in $\mathcal{D}'(I, X)$. Then $f = 0$ almost everywhere.

Proof. First, we remark that if J is a bounded subinterval of I, we have $\int_J f = 0$. Indeed, let $(\varphi_n)_{n \geq 1} \subset \mathcal{D}(I)$, $\varphi_n \leq 1$, and $\varphi_n \to 1_J$ almost everywhere. We have
$$\int_J f = \lim \int f \varphi_n = \lim \langle f, \varphi_n \rangle = 0.$$
Then fix a bounded subinterval $J \subset I$ and consider $\overline{f} \in L^1(\mathbb{R}, X)$, defined by
$$\begin{cases} \overline{f}(t) = f(t), & \text{if } t \in J, \\ \overline{f}(t) = 0, & \text{if } t \notin J. \end{cases}$$
It follows that $T_h \overline{f} = 0$ for all $h > 0$. By Proposition 1.4.29, we obtain $\overline{f} = 0$ almost everywhere. Therefore, $f = 0$ almost everywhere on J. Since J is arbitrary, we have $f = 0$ almost everywhere. □

Corollary 1.4.31. *Let $g \in L^1_{\text{loc}}(I, X)$, $t_0 \in I$ and let $f \in C(I, X)$ be given by $f(t) = \int_{t_0}^t g(s) \, ds$. Then:*

(i) *$f' = g$ in $\mathcal{D}'(I, X)$;*

(ii) *f is differentiable almost everywhere and $f' = g$ almost everywhere.*

Proof. Reasoning as before, we may restrict ourselves to the case $I = \mathbb{R}$ and $g \in L^1(\mathbb{R}, X)$. We have
$$T_h g(t) = \frac{f(t+h) - f(t)}{h}.$$
By Proposition 1.4.29, we deduce (ii).

Now consider $\varphi \in \mathcal{D}(\mathbb{R})$. We have
$$\langle f', \varphi \rangle = -\langle f, \varphi' \rangle = -\int f(t) \varphi'(t) \, dt.$$
Furthermore,
$$\frac{\varphi(t+h) - \varphi(t)}{h} \longrightarrow \varphi'(t) \quad \text{uniformly on } \mathbb{R}, \text{ as } h \to 0.$$
Therefore,
$$\langle f', \varphi \rangle = -\lim_{h \to 0} \int_{\mathbb{R}} f(t) \frac{\varphi(t+h) - \varphi(t)}{h}$$
$$= -\lim_{h \to 0} \int_{\mathbb{R}} \varphi(t) \frac{f(t-h) - f(t)}{h} = \lim_{h \to 0} \int_{\mathbb{R}} T_{-h} g \varphi = \langle g, \varphi \rangle,$$
by Proposition 1.4.29; hence (i). □

Proposition 1.4.32. Let $T \in \mathcal{D}'(I, X)$ be such that $T' = 0$. Then there exists $x_0 \in X$ such that $T = x_0$, i.e.

$$\langle T, \varphi \rangle = x_0 \int_I \varphi,$$

for all $\varphi \in \mathcal{D}(I)$.

Proof. Let $\theta \in \mathcal{D}(I)$ be such that $\int_I \theta = 1$, and let $x_0 = \langle T, \theta \rangle$. Let (a, b) be the support of θ and let $t_0 \in I$, $t_0 < a$. Now consider $\varphi \in \mathcal{D}(I)$. We define $\psi \in \mathcal{D}(I)$ by

$$\psi(t) = \int_{t_0}^t \left(\varphi(s) - \theta(s) \int_I \varphi(\sigma) \, d\sigma \right) ds,$$

for all $t \in I$. We have

$$\psi' = \varphi - \theta \left(\int_I \varphi \right).$$

Hence

$$0 = \langle T, \psi' \rangle = \langle T, \varphi \rangle - x_0 \int_I \varphi.$$

It follows that

$$\langle T, \varphi \rangle = x_0 \int_I \varphi;$$

hence the result. $\square$

1.4.5. The spaces $W^{1,p}(I, X)$

Definition 1.4.33. Let $1 \le p \le \infty$. We denote by $W^{1,p}(I, X)$ the space of (equivalence classes of) functions $f \in L^p(I, X)$ such that $f' \in L^p(I, X)$, in the sense of $\mathcal{D}'(I, X)$. For $f \in W^{1,p}(I, X)$, we set $\|f\|_{W^{1,p}} = \|f\|_{L^p} + \|f'\|_{L^p}$.

Proposition 1.4.34. $(W^{1,p}(I, X), \|\cdot\|_{W^{1,p}})$ is a Banach space.

Proof. The proof is similar to that of the real case. $\square$

Theorem 1.4.35. Let $1 \le p \le \infty$ and let $f \in L^p(I, X)$. Then the following properties are equivalent:

(i) $f \in W^{1,p}(I, X)$;

(ii) there exists $g \in L^p(I, X)$ such that for almost all $t_0, t \in I$, we have $f(t) = f(t_0) + \int_{t_0}^t g(s) \, ds$;

(iii) there exists $g \in L^p(I, X)$, $x_0 \in X$, and $t_0 \in I$ such that we have $f(t) = x_0 + \int_{t_0}^t g(s) \, ds$, for almost all $t \in I$;

(iv) f is absolutely continuous, differentiable almost everywhere, and f' belongs to $L^p(I,X)$;

(v) f is weakly absolutely continuous, weakly differentiable almost everywhere, and f' belongs to $L^p(I,X)$.

Proof. (i)$\Rightarrow$(ii). Let $t_0 \in I$. For any $t \in I$, set

$$w(t) = f(t) - f(t_0) - \int_{t_0}^{t} f'(s)\,ds.$$

We have $w \in C(I,X)$ and $w' = 0$ in $\mathcal{D}'(I,X)$ by Corollary 1.4.31. Therefore, there exists $x_0 \in X$ such that $w = x_0$ in $\mathcal{D}'(I,X)$ (Proposition 1.4.32). Since $w(0) = 0$, it follows from Corollary 1.4.30 that $w = 0$ almost everywhere; hence (ii).

(ii)$\Rightarrow$(iii) is immediate.

(iii)$\Rightarrow$(iv). By possibly modifying f on a set of measure 0, we may assume that

$$f(t) = x_0 + \int_{t_0}^{t} g(s)\,ds,$$

for all $t \in I$, and we apply Corollary 1.4.31.

(iv)$\Rightarrow$(v) is immediate.

(v)$\Rightarrow$(i). Let g be the weak derivative of f. Let $t_0 \in I$, and set

$$\varphi(t) = f(t) - f(t_0) - \int_{t_0}^{t} g(s)\,ds,$$

for all $t \in I$. By Corollary 1.4.31, φ is differentiable almost everywhere and its derivative is 0 almost everywhere.

Let $x' \in X^*$ and let ψ be defined by $\psi(t) = \langle x', \varphi(t) \rangle$. ψ is absolutely continuous, differentiable almost everywhere and $\psi'(t) = 0$ almost everywhere. Since $\psi(t_0) = 0$, we obtain $\psi(t) \equiv 0$. Since x' is arbitrary, we conclude that $\varphi(t) \equiv 0$. Hence, (i) follows from Corollary 1.4.31. $\square$

Corollary 1.4.36. Let $1 \le p \le \infty$. Then $W^{1,p}(I,X) \hookrightarrow C_{b,u}(I,X)$.

Proof. We have

$$\|f(t) - f(s)\| \le \int_{s}^{t} \|f'(\sigma)\|\,d\sigma,$$

for all $s,t \in I$. Hence we have uniform continuity.

Furthermore, if we set $h(\cdot) = \|f(\cdot)\|$, we have

$$|h(t) - h(s)| \le \|f(t) - f(s)\|.$$

It follows that
$$|h(t) - h(s)| \leq \int_s^t \|f'(\sigma)\| \, d\sigma,$$
for all $s, t \in I$. Therefore, h is absolutely continuous and we have $|h'| \leq \|f'\| \in L^p(I)$ almost everywhere. We obtain $h \in W^{1,p}(I) \hookrightarrow L^\infty(I)$, which completes the proof. □

Corollary 1.4.37. *If I is bounded, then $C^\infty(\overline{I}, X)$ is dense in $W^{1,p}(I, X)$.*

Proof. Let $f \in W^{1,p}(I, X)$ and let $(g_n)_{n \geq 1} \subset \mathcal{D}(I, X)$ be such that $g_n \to f'$ in $L^p(I, X)$. Let $t_0 \in I$, and set
$$f_n(t) = f(t_0) + \int_{t_0}^t g_n(s) \, ds.$$
It is now easy to verify that $f_n \in C^\infty(\overline{I}, X)$ and that $f_n \longrightarrow f$ in $W^{1,p}(I, X)$, as $n \to \infty$. □

Corollary 1.4.38. *Let $1 < p < \infty$. Then $W^{1,p}(I, X) \hookrightarrow C^{0,\alpha}(I, X)$, with $\alpha = (p-1)/p$.*

Proof. By Hölder's inequality, we have
$$\|f(t+h) - f(t)\| \leq h^{\frac{1}{p'}} \left(\int_t^{t+h} \|f'(s)\|^p \, ds \right)^{\frac{1}{p}} \leq h^{\frac{1}{p'}} \|f'\|_{L^p};$$
hence the result. □

Corollary 1.4.39. *Assume that $I = (a, b)$. Let $1 \leq p < \infty$ and let $f \in W^{1,p}(I, X)$. Then, for all $c \in I$, we have*
$$\frac{f(\cdot + h) - f(\cdot)}{h} \longrightarrow f'$$
in $L^p((a, c), X)$, as $h \downarrow 0$.

Proof. We extend f on $\mathbb{R}$ by a function $\overline{f} \in W^{1,p}(\mathbb{R}, X)$ and we apply Proposition 1.4.28 and Theorem 1.4.35. □

Theorem 1.4.40. *Suppose that X is reflexive. Let $1 \leq p \leq \infty$ and let $f \in L^p(I, X)$. Then $f \in W^{1,p}(I, X)$ if and only if there exists $\varphi \in L^p(I, \mathbb{R})$ such that*
$$\|f(\tau) - f(t)\| \leq \left| \int_t^\tau \varphi(s) \, ds \right|, \qquad (1.1)$$
for almost all $t, \tau \in I$. In that case, we have $\|f'\|_{L^p(I, X)} \leq \|\varphi\|_{L^p(I, \mathbb{R})}$.

Proof. It is clear that this condition is necessary (for example, take $\varphi(\cdot) = \|f'(\cdot)\|$).

Conversely, since X is complete, we easily verify that we can modify f in a set of measure 0 so that (1.1) holds for all $t, \tau \in I$. In particular, f is continuous and we may consider only the case in which $I = \mathbb{R}$. Since f is continuous, it is clear that $f(\mathbb{R})$ is separable. By possibly considering the smallest closed subspace of X, we restrict ourselves to the case in which X is reflexive and separable, and so X^* is separable. For all $h > 0$, set

$$f_h(t) = \frac{f(t+h) - f(t)}{h}.$$

If $p = \infty$, it is clear that f_h is bounded in $L^p(\mathbb{R}, X)$. If $p < \infty$, by Hölder's inequality, we have

$$\|f_h(t)\|^p \le \frac{1}{h} \int_t^{t+h} |\varphi(s)|^p \, ds.$$

Integrating on $\mathbb{R}$, we find that

$$\int_{\mathbb{R}} \|f_h(t)\|^p \le \frac{1}{h} \int_{\mathbb{R}} \int_t^{t+h} |\varphi(s)|^p \, ds \, dt = \frac{1}{h} \int_{\mathbb{R}} \int_{s-h}^{s} |\varphi(t)|^p \, dt \, ds \qquad (1.2)$$
$$= \int_{\mathbb{R}} |\varphi(s)|^p \, ds.$$

Therefore f_h is bounded in $L^p(\mathbb{R}, X)$. Let (x'_n) be dense in X^*. For all n, the function $\psi_n(t) = \langle x'_n, f(t) \rangle$ satisfies

$$|\psi_n(\tau) - \psi_n(t)| \le \|x'_n\| \left| \int_t^\tau \varphi(s) \, ds \right|.$$

In particular, ψ_n is absolutely continuous, and so differentiable on $\mathbb{R} \setminus E_n$, with $|E_n| = 0$. Let $E = \bigcup_{n \ge 1} E_n$. We have $|E| = 0$, and for all $t \in \mathbb{R} \setminus E$ and $n \in \mathbb{N}$, we have $\langle f_h(t), x'_n \rangle \to \psi'_n(t)$, as $h \to 0$. Let F be the complement of the set of Lebesgue's points of φ (we know $|F| = 0$). By (1.1), for all $t \in \mathbb{R} \setminus F$, we have $\|f_h(t)\| \le K(t) < \infty$, if $|h|$ is small enough. We claim that for all $t \in \mathbb{R} \setminus (E \cup F)$, there exists $w(t) \in X$ such that $f_h(t) \rightharpoonup w(t)$ as $h \to 0$. Indeed, $\|f_h(t)\|$ is bounded, and since X is reflexive, there exists a sequence $h_n \to 0$ and $w(t) \in X$, such that $f_{h_n}(t) \rightharpoonup w(t)$ weakly in X as $n \to \infty$.

In particular, we have $\langle w(t), x'_n \rangle = \psi'_n(t)$, for all $n \in \mathbb{N}$. Since the sequence $(x'_n)_{n \ge 1}$ is dense in X^*, $w(t)$ does not depend on the sequence h_n, and so $f_h(t) \rightharpoonup w(t)$. By Proposition 1.4.24, we have $w \in L^p(\mathbb{R}, X)$, and $\|w\|_{L^p(I,X)} \le \|\varphi\|_{L^p(I,\mathbb{R})}$. By Theorem 1.4.35(v) and Corollary 1.4.31, we have $f \in W^{1,p}(\mathbb{R}, X)$ and $f' = w$. $\square$

From this, we immediately deduce the following result.

Corollary 1.4.41. *Assume that X is reflexive. Let $f : I \to X$ be bounded, Lipschitz continuous with Lipschitz constant L. Then, $f \in W^{1,\infty}(I, X)$ and $\|f'\|_{L^\infty(I,X)} \leq L$.*

Corollary 1.4.42. *Assume that X is reflexive and that $1 < p \leq \infty$. Let $(f_n)_{n \geq 0}$ be a bounded sequence in $W^{1,p}(I, X)$ and let $f : I \to X$ be such that $f_n(t) \rightharpoonup f(t)$ as $n \to \infty$, for almost every $t \in I$. Then $f \in W^{1,p}(I, X)$, and $\|f'\|_{L^p(I,X)} \leq \liminf\limits_{n \to \infty} \|f_n'\|_{L^p(I,X)}$.*

Proof. By Proposition 1.4.24, we have $f \in L^p(I, X)$. Let E be a set of measure 0 such that $f_n(t) \rightharpoonup f(t)$ as $n \to \infty$, for all $t \in I \setminus E$. For all $t, \tau \in I \setminus E$, we have

$$\|f(t) - f(\tau)\| \leq \liminf_{n \to \infty} \|f_n(t) - f_n(\tau)\| \leq \liminf_{n \to \infty} \int_t^\tau \|f_n'(s)\| \, ds. \tag{1.3}$$

Consider $\varphi_n = \|f_n'\|$. φ_n is bounded in $L^p(I)$ and so there exists a subsequence $(\varphi_{n_k})_{k \geq 1}$ and $\varphi \in L^p(I)$ such that $\varphi_{n_k} \rightharpoonup \varphi$ weakly in $L^p(I)$ (weak-$\star$ if $p = \infty$) as $k \to \infty$, and $\liminf\limits_{k \to \infty} \|\varphi_{n_k}\|_{L^p} = \liminf\limits_{n \to \infty} \|\varphi_n\|_{L^p}$. In particular, we have

$$\int_t^\tau \varphi_{n_k}(s) \, ds \longrightarrow \int_t^\tau \varphi(s) \, ds, \quad \text{as } k \to \infty, \tag{1.4}$$

for all $t, \tau \in I$, and

$$\|\varphi\|_{L^p} \leq \liminf_{n \to \infty} \|f_n'\|_{L^p}. \tag{1.5}$$

It follows from (1.3) and (1.4) that

$$\|f(t) - f(\tau)\| \leq \int_t^\tau \varphi(s) \, ds,$$

for all $t, \tau \in I$; therefore $f \in W^{1,p}(I, X)$ (Theorem 1.4.40). We complete the proof by applying (1.5). □

Notes. For §1.1 and §1.2, consult Brezis [2], Dunford and Schwartz [2], and Yosida [1]. For §1.3, see Adams [1], Bergh and Löfstrom [1], Brezis [2], and Gilbarg and Trudinger [1], and for §1.4, see Dinculeanu [1], Dunford and Schwartz [1], and the appendix of Brezis [1].

2

m-dissipative operators

Throughout this chapter, X is a Banach space, endowed with the norm $\|\cdot\|$.

2.1. Unbounded operators in Banach spaces

Definition 2.1.1. A linear unbounded operator in X is a pair (D, A), where D is a linear subspace of X and A is a linear mapping $D \to X$. We say that A is bounded if there exists $c > 0$ such that

$$\|Au\| \leq c,$$

for all $u \in \{x \in D, \|x\| \leq 1\}$. Otherwise, A is not bounded.

Remark 2.1.2. Note that a linear unbounded operator can be either bounded or not bounded. This somewhat strange terminology is in general use and should not lead to misunderstanding in our applications.

Remark 2.1.3. If A is bounded, A is the restriction to D of an operator $\widetilde{A} \in \mathcal{L}(Y, X)$, where Y is a closed linear subspace of X, such that $D \subset Y$. If A is not bounded, there exists no operator $\widetilde{A} \in \mathcal{L}(Y, X)$ with Y closed in X and $D \subset Y$, such that $\widetilde{A}_{|D} = A$.

Definition 2.1.4. Let (D, A) be a linear operator in X. The graph $G(A)$ of A and the range $R(A)$ of A are defined by

$$G(A) = \{(u, f) \in X \times X; u \in D \text{ and } f = Au\},$$
$$R(A) = A(D).$$

$G(A)$ is a linear subspace of $X \times X$, and $R(A)$ is a linear subspace of X.

Remark 2.1.5. In this chapter, a linear unbounded operator is just called an operator where there is no risk of confusion. As usual, we denote the pair (D, A) by A with $D(A) = D$, meaning the domain of A is D. Note, however, that when one defines an operator, it is absolutely necessary to define its domain.

Remark 2.1.6. When $D(A) = X$, it follows from Theorem 1.1.2 that $A \in \mathcal{L}(X)$ if and only if $G(A)$ is closed in X. More generally, for not bounded operators, it is very useful to know whether or not the graph is closed.

2.2. Definition and main properties of m-dissipative operators

Definition 2.2.1. An operator A in X is dissipative if

$$\|u - \lambda Au\| \geq \|u\|,$$

for all $u \in D(A)$ and all $\lambda > 0$.

Definition 2.2.2. An operator A in X is m-dissipative if

(i) A is dissipative;

(ii) for all $\lambda > 0$ and all $f \in X$, there exists $u \in D(A)$ such that $u - \lambda Au = f$.

Remark 2.2.3. If A is m-dissipative in X, it is clear, from Definitions 2.2.1 and 2.2.2, that for all $f \in X$ and all $\lambda > 0$, there exists a unique solution u of the equation $u - \lambda Au = f$. In addition, one has $\|u\| \leq \|f\|$.

Definition 2.2.4. Let A be an m-dissipative operator in X and $\lambda > 0$. For all $f \in X$, we denote by $J_\lambda j$ or by $(I - \lambda A)^{-1} f$ the solution u of the equation $u - \lambda Au = f$.

Remark 2.2.5. By Remark 2.2.3, one has $J_\lambda \in \mathcal{L}(X)$ and $\|J_\lambda\|_{\mathcal{L}(X)} \leq 1$.

Proposition 2.2.6. Let A be a dissipative operator in X. The following properties are equivalent.

(i) A is m-dissipative in X;

(ii) there exists $\lambda_0 > 0$ such that for all $f \in X$, there exists a solution $u \in D(A)$ of $u - \lambda_0 Au = f$.

Proof. It is clear that (i)$\Rightarrow$(ii). Let us show that (ii)$\Rightarrow$(i). Let $\lambda > 0$. Note that the equation $u - \lambda Au = f$ is equivalent to

$$u - \lambda_0 Au = \frac{\lambda_0}{\lambda} f + \left(1 - \frac{\lambda_0}{\lambda}\right) u.$$

Since A is dissipative and $R(I - \lambda_0 A) = X$, the operator $J_{\lambda_0} = (I - \lambda_0 A)^{-1}$ can be defined as in Definition 2.2.4. This operator is a contraction on X. Next, note that the preceding equation is also equivalent to

$$u = J_{\lambda_0} \left(\frac{\lambda_0}{\lambda} f + \left(1 - \frac{\lambda_0}{\lambda}\right) u\right).$$

If $2\lambda > \lambda_0$, this last equation is $u = F(u)$, where F is Lipschitz continuous $x \longrightarrow X$, with a Lipschitz constant $k = |(\lambda - \lambda_0)/\lambda| < 1$. Applying Theorem 1.1.1, there exists a solution u of $u - \lambda Au = f$, for all $\lambda \in (\lambda_0/2, \infty)$.

Iterating this argument n times, there exists a solution for all $\lambda \in (2^{-n}\lambda_0, \infty)$, $n \geq 1$. Since n is arbitrary there exists a solution for all $\lambda > 0$. □

Proposition 2.2.7. *If A is m-dissipative, then $G(A)$ is closed in X.*

Proof. Since $J_1 \in \mathcal{L}(X)$, $G(J_1)$ is closed. It follows that $G(I - A)$ is closed, and so $G(A)$ is closed. □

Corollary 2.2.8. *Let A be an m-dissipative operator. For every $u \in D(A)$, let $\|u\|_{D(A)} = \|u\| + \|Au\|$. Then $(D(A), \|\cdot\|_{D(A)})$ is a Banach space, and $A \in \mathcal{L}(D(A), X)$.*

Remark 2.2.9. In what follows, and in particular in Chapters 3 and 4, $D(A)$ means the Banach space $(D(A), \|\cdot\|_{D(A)})$.

Proposition 2.2.10. *If A is m-dissipative, then $\lim_{\lambda \downarrow 0} \|J_\lambda u - u\| = 0$ for all $u \in \overline{D(A)}$.*

Proof. We have $\|J_\lambda - I\| \leq 2$, and by density we need only consider the case $u \in D(A)$. We have
$$J_\lambda u - u = J_\lambda (u - (I - \lambda A)u);$$
and so $\|J_\lambda u - u\| \leq \|u - (I - \lambda A)u\| = \lambda \|Au\| \to 0$, as $\lambda \downarrow 0$. □

Definition 2.2.11. Let A be an m-dissipative operator. For $\lambda > 0$, we denote by A_λ the operator defined by
$$A_\lambda = AJ_\lambda = \frac{J_\lambda - I}{\lambda}.$$
We have $A_\lambda \in \mathcal{L}(X)$ and $\|A_\lambda\|_{\mathcal{L}(X)} \leq 2/\lambda$.

Proposition 2.2.12. *If A is m-dissipative and if $\overline{D(A)} = X$, then $A_\lambda u \longrightarrow Au$ as $\lambda \downarrow 0$ for all $u \in D(A)$.*

Proof. Let $u \in D(A)$. By Proposition 2.2.10, one has
$$J_\lambda Au - Au \longrightarrow 0 \quad \text{as } \lambda \downarrow 0.$$
On the other hand, it follows easily from Definition 2.2.11 that
$$A_\lambda u = J_\lambda Au.$$
Thus,
$$\|A_\lambda u - Au\| = \|J_\lambda Au - Au\| \longrightarrow 0 \quad \text{as } \lambda \downarrow 0;$$
hence the result. □

2.3. Extrapolation

In this section, we show that, given an m-dissipative operator A on X with dense domain, one can extend it to an m-dissipative operator $\overline{A}$ on a larger space $\overline{X}$. This result will be very useful for characterizing the weak solutions in Chapters 3 and 4.

Proposition 2.3.1. *Let A be an m-dissipative operator in X with dense domain. There exists a Banach space $\overline{X}$, and an m-dissipative operator $\overline{A}$ in $\overline{X}$, such that*

(i) $X \hookrightarrow \overline{X}$, with dense embedding;

(ii) for all $u \in X$, the norm of u in $\overline{X}$ is equal to $\|J_1 u\|$;

(iii) $D(\overline{A}) = X$, with equivalent norms;

(iv) $\overline{A}u = Au$, for all $u \in D(A)$.

In addition, $\overline{X}$ and $\overline{A}$ satisfying (i)–(iv) are unique, up to isomorphism.

Proof. For $u \in X$, we define $\|\|u\|\| = \|J_1 u\|$. It is clear that $\|\| \cdot \|\|$ is a norm on X. Let $\overline{X}$ be the completion of X for the norm $\|\| \cdot \|\|$. $\overline{X}$ is unique, up to an isomorphism, and $X \hookrightarrow \overline{X}$, with dense embedding. On the other hand, observe that
$$J_1 A u = J_1 u - u, \quad \forall u \in D(A).$$
Thus,
$$\|\|Au\|\| \le \|\|u\|\| + \|u\| \le 2\|u\|, \quad \forall u \in D(A).$$
Hence, A can be extended to an operator $\widetilde{A} \in \mathcal{L}(X, Y)$. We define the linear operator $\overline{A}$ on $\overline{X}$ by
$$D(\overline{A}) = X,$$
$$\overline{A}u = \widetilde{A}u, \quad \forall u \in D(\overline{A}).$$
It is clear that $\overline{A}$ satisfies (iii) and (iv). Now, let us show that $\overline{A}$ is dissipative. Take $\lambda > 0$. Let $u \in D(A)$ and let $v = J_1 u$. One has
$$v - \lambda Av = J_1(u - \lambda Au).$$
Since A is dissipative, it follows that
$$\|\|u - \lambda Au\|\| = \|v - \lambda Av\| \ge \|v\| = \|\|u\|\|.$$
By continuity of $\widetilde{A}$, we deduce that
$$\|\|u - \lambda \overline{A}u\|\| \ge \|\|u\|\|, \quad \forall u \in X;$$

and so $\overline{A}$ is dissipative. Finally, let $f \in \overline{X}$, and $(f_n)_{n \geq 0} \subset X$, with $f_n \longrightarrow f$ in $\overline{X}$ as $n \to \infty$. Set $u_n = J_1 f_n$. Since $(f_n)_{n \geq 0}$ is a Cauchy sequence in $\overline{X}$, $(u_n)_{n \geq 0}$ is also a Cauchy sequence in X; and so there exists $u \in X$, such that $u_n \longrightarrow u$ in X as $n \to \infty$. We have

$$f_n = u_n - Au_n = u_n - \tilde{A}u_n.$$

Since $\tilde{A} \in \mathcal{L}(X,Y)$; it follows that $f = u - \tilde{A}u = u - \overline{A}u$. Hence $\overline{A}$ is m-dissipative. The uniqueness of $\overline{A}$ follows from the uniqueness of $\tilde{A}$. □

Corollary 2.3.2. *If $x \in X$ is such that $\overline{A}x \in X$, then $x \in D(A)$ and $Ax = \overline{A}x$.*

Proof. Let $f = x - \overline{A}x \in X$. Since A is m-dissipative, there exists $y \in D(A)$ such that $y - Ay = f$. By Proposition 2.3.1(iii), we have $(x-y) - \overline{A}(x-y) = 0$, and since $\overline{A}$ is dissipative, we obtain $x = y$. □

2.4. Unbounded operators in Hilbert spaces

Throughout this section, we assume that X is a Hilbert space, and we denote by $\langle \cdot, \cdot \rangle$ its scalar product. If A is a linear operator in X with dense domain, then

$$G(A^*) = \{(v,\varphi) \in X \times X; \langle \varphi, u \rangle = \langle v, f \rangle \text{ for all } (u,f) \in G(A)\},$$

defines a linear operator A^* (the adjoint of A). The domain of A^* is

$$D(A^*) = \{v \in X, \exists C < \infty, |\langle Au, v \rangle| \leq C\|u\|, \forall u \in D(A)\},$$

and A^* satisfies

$$\langle A^*v, u \rangle = \langle v, Au \rangle, \forall u \in D(A),$$

Indeed, the linear mapping $u \mapsto \langle v, Au \rangle$, defined on $D(A)$ for all $v \in D(A^*)$, can be extended to a unique linear mapping $\varphi \in X' \approx X$, denoted by $\varphi = A^*v$. It is clear that $G(A^*)$ is systematically closed.

Finally, it follows easily that if $B \in \mathcal{L}(X)$, then $(A+B)^* = A^* + B^*$.

Proposition 2.4.1. $\left(\overline{R(A)}\right)^\perp = \{v \in D(A^*); A^*v = 0\}$.

Proof. One has $v \in \left(\overline{R(A)}\right)^\perp \Leftrightarrow \langle v, Au \rangle = 0, \forall u \in D(A) \Leftrightarrow (0,v) \in G(A^*)$. This last property is equivalent to $v \in D(A^*)$ and $A^*v = 0$; hence the result. □

Proposition 2.4.2. *A is dissipative in X if and only if $\langle Au, u \rangle \leq 0$, for all $u \in D(A)$.*

Proof. If A is dissipative, one has

$$-2\lambda \langle Au, u \rangle + \lambda^2 \|Au\|^2 = \|u - \lambda Au\|^2 - \|u\|^2 \geq 0, \quad \forall \lambda > 0, \forall u \in D(A).$$

Dividing by λ and letting $\lambda \downarrow 0$, we obtain

$$\langle Au, u \rangle \leq 0, \text{ for all } u \in D(A).$$

Conversely, if the last property is satisfied, then for all $\lambda > 0$ and $u \in D(A)$ we have

$$\|u - \lambda Au\|^2 = \|u\|^2 - 2\lambda \langle Au, u \rangle + \lambda^2 \|Au\|^2 \geq \|u\|^2,$$

and then A is dissipative. □

Corollary 2.4.3. *If A is m-dissipative in X, then $D(A)$ is dense in X.*

Proof. Let $z \in (D(A))^\perp$, and let $u = J_1 z \in D(A)$. We have

$$0 = \langle z, u \rangle = \langle u - Au, u \rangle.$$

Hence,
$$\|u\|^2 = \langle Au, u \rangle \leq 0.$$

It follows that $u = z = 0$; and so $D(A)$ is dense in X. □

Corollary 2.4.4. *If A is m-dissipative in X, then*

$$J_\lambda u \longrightarrow u \quad \text{as } \lambda \downarrow 0,$$

for all $u \in X$ and

$$A_\lambda u \longrightarrow Au \quad \text{as } \lambda \downarrow 0,$$

for all $u \in D(A)$.

Proof. We apply Corollary 2.2.3 and Propositions 2.2.10 and 2.2.12. □

Theorem 2.4.5. *Let A be a linear dissipative operator in X with dense domain. Then A is m-dissipative if and only if A^* is dissipative and $G(A)$ is closed.*

Proof. If A is m-dissipative, then $G(A)$ is closed, by Proposition 2.2.7. Let us show that A^* is dissipative. Let $v \in D(A^*)$. We have

$$\langle A^* v, J_\lambda v \rangle = \langle v, A J_\lambda v \rangle = \langle v, A_\lambda v \rangle$$
$$= \frac{1}{\lambda} \langle v, J_\lambda v - v \rangle = \frac{1}{\lambda} \{ \langle v, J_\lambda v \rangle - \|v\|^2 \} \leq 0.$$

Since $\langle A^*v, J_\lambda v\rangle \longrightarrow \langle A^*v, v\rangle$ as $\lambda \downarrow 0$, it follows that A^* is dissipative.

Conversely, since A is dissipative and $G(A)$ is closed, it is clear that $R(I-A)$ is closed in X. On the other hand, by Proposition 2.4.1, one has

$$\left(\overline{R(I-A)}\right)^\perp = \{v \in D(A^*); v - A^*v = 0\} = \{0\},$$

since A^* is dissipative. Therefore $R(I-A) = X$, and A is m-dissipative, by Proposition 2.2.6. □

Definition 2.4.6. Let A be a linear operator in X with dense domain. We say that A is self-adjoint (respectively skew-adjoint) if $A^* = A$ (respectively $A^* = -A$).

Remark 2.4.7. The equality $A^* = \pm A$ has to be taken in the sense of operators. It means that $D(A) = D(A^*)$ and $A^*u = \pm Au$, for all $u \in D(A)$.

Corollary 2.4.8. If A is a self-adjoint operator in X, and if $A \leq 0$ (i.e. $\langle Au, u\rangle \leq 0$, for all $u \in D(A)$), then A is m-dissipative.

Proof. By Proposition 2.4.2, A is dissipative. Since $A^* = A$, A^* is dissipative. Finally, $G(A^*)$ is closed, so that $G(A)$ is closed. We finish the proof by applying Theorem 2.4.5. □

Corollary 2.4.9. If A is a skew-adjoint operator in X, then A and $-A$ are m-dissipative.

Proof. Let $u \in D(A)$. One has $\langle Au, u\rangle = \langle u, A^*u\rangle = -\langle u, Au\rangle$. Hence $\langle Au, u\rangle = 0$. It follows from Proposition 2.4.2, that A and $-A$ are dissipative. We conclude as in Corollary 2.4.8. □

Corollary 2.4.10. Let A be a linear operator in X with dense domain, such that $G(A) \subset G(A^*)$ and $A \leq 0$. Then A is m-dissipative if and only if A is self-adjoint.

Proof. Applying Corollary 2.4.8, we need only show that if A is m-dissipative then A is self-adjoint. Let $(u, f) \in G(A^*)$, and let $g = u - A^*u = u - f$. Since A is m-dissipative, there exists $v \in D(A)$ such that $g = v - Av$, and since $G(A) \subset G(A^*)$, we have $v \in D(A^*)$ and $g = v - A^*v$. Therefore $(v-u) - A^*(v-u) = 0$ and since A^* is dissipative (Theorem 2.4.5), we obtain $u = v$. Thus, $(u, f) \in G(A)$; and so $A = A^*$. □

Corollary 2.4.11. *Let A be a linear operator in X with dense domain. Then A and $-A$ are m-dissipative if and only if A is skew-adjoint.*

Proof. Applying Corollary 2.4.9, it suffices to show that if A and $-A$ are m-dissipative, then A is skew-adjoint. Applying Proposition 2.4.2 to A and $-A$, we obtain
$$\langle Au, u\rangle = 0, \quad \forall u \in D(A).$$
For all $u, v \in D(A)$, we obtain
$$\langle Au, v\rangle + \langle Av, u\rangle = \langle A(u+v), u+v\rangle - \langle Au, u\rangle - \langle Av, v\rangle = 0.$$
Therefore $G(-A) \subset G(A^*)$. It remains to show that $G(A^*) \subset G(-A)$. Consider $(u, f) \in G(A^*)$ and let $g = u - A^*u = u - f$. Since $-A$ is m-dissipative, there exists $v \in D(A)$ such that $g = v + Av$, and since $G(-A) \subset G(A^*)$, we have $v \in D(A^*)$ and $f = v - A^*v$. Hence $(v-u) - A^*(v-u) = 0$ and since $-A^*$ is dissipative (Theorem 2.4.5), we obtain $u = v$. Therefore, $(u, f) \in G(A^*)$; and so $A = -A^*$. □

2.5. Complex Hilbert spaces

In this section, we assume that X is a complex Hilbert space. Recall that by definition X is a complex Hilbert space provided that there exists a continuous $\mathbb{R}$-bilinear mapping $b : X \times X \to \mathbb{C}$ satisfying the following properties:
$$\begin{aligned}b(iu, v) &= ib(u, v), \quad \forall(u, v) \in X \times X;\\ b(v, u) &= \overline{b(u, v)}, \quad \forall(u, v) \in X \times X;\\ b(u, u) &= \|u\|^2, \quad \forall u \in X.\end{aligned}$$
In that case $\langle u, v\rangle = \operatorname{Re}(b(u, v))$ defines a (real) scalar product on X. Equipped with this scalar product, X is a real Hilbert space. In what follows, we consider X as a real Hilbert space.

Let A be a linear operator on the real Hilbert space X. If A is $\mathbb{C}$-linear, we can define iA as a linear operator on the real Hilbert space X.

Proposition 2.5.1. *Assume that $D(A)$ is dense and that A is $\mathbb{C}$-linear. Then A^* is $\mathbb{C}$-linear, and $(iA)^* = -iA^*$.*

Proof. Let $v \in D(A)$, $f = A^*v$ and let $z \in \mathbb{C}$. For all $u \in D(A)$, we have
$$\langle zf, u\rangle = \langle f, \overline{z}u\rangle = \langle v, A(\overline{z}u)\rangle = \langle v, \overline{z}Au\rangle = \langle zv, Au\rangle.$$
Therefore $zv \in D(A^*)$ and $zf = A(zv)$. Hence A^* is $\mathbb{C}$-linear. In addition,
$$\langle -if, u\rangle = \langle v, A(iu)\rangle = \langle v, iAu\rangle,$$

for all $(v,f) \in G(A^*)$ and all $u \in D(A)$; and so $G(-iA^*) \subset G((iA)^*)$. Applying this result to iA, we obtain

$$G(-i(iA)^*) \subset G((i \cdot iA)^*) = G(-A^*).$$

It follows that $G((iA)^*) \subset G(-iA^*)$, and so $G((iA)^*) = G(-iA^*)$. □

Corollary 2.5.2. *If A is self-adjoint, then iA is skew-adjoint.*

Proof. $(iA)^* = -iA^* = -iA$. □

2.6. Examples in the theory of partial differential equations

2.6.1. The Laplacian in an open subset of $\mathbb{R}^N$: L^2 theory

Let Ω be any open subset of $\mathbb{R}^N$, and let $Y = L^2(\Omega)$. We can consider either real-valued functions or complex-valued functions, but in both cases, Y is considered as a real Hilbert space (see §2.5). We define the linear operator B in Y by

$$\begin{cases} D(B) = \{u \in H_0^1(\Omega); \, \Delta u \in L^2(\Omega)\}; \\ Bu = \Delta u, \quad \forall u \in D(B). \end{cases}$$

Proposition 2.6.1. *B is m-dissipative with dense domain. More precisely, B is self-adjoint and $B \leq 0$.*

We need the following lemma.

Lemma 2.6.2. *We have*

$$\int_\Omega v \Delta u \, dx = -\int_\Omega \nabla u \cdot \nabla v \, dx. \tag{2.1}$$

for all $u \in D(B)$ and $v \in H_0^1(\Omega)$.

Proof. (2.1) is satisfied by $v \in \mathcal{D}(\Omega)$. The lemma follows by density, since both terms of (2.1) are continuous in v on $H_0^1(\Omega)$. □

Proof of Proposition 2.6.1. First, $\mathcal{D}(\Omega) \subset D(B)$, and so $D(B)$ is dense in Y. Let $u \in D(B)$. Applying (2.1) with $v = u$, we obtain $\langle Bu, u \rangle \leq 0$, so that B is dissipative (Proposition 2.4.2). The bilinear continuous mapping

$$b(u,v) = \int (uv + \nabla u \cdot \nabla v) dx$$

is coercive in $H_0^1(\Omega)$. It follows from Theorem 1.1.4 that, for all $f \in L^2(\Omega)$, there exists $u \in H_0^1(\Omega)$ such that

$$\int (uv + \nabla u \cdot \nabla v)\,dx = \int fv\,dx, \quad \forall v \in H_0^1(\Omega).$$

We obtain

$$u - \Delta u = f,$$

in the sense of distributions. Since, in addition $u \in H_0^1(\Omega)$, we obtain $u \in D(B)$ and $u - Bu = f$. Therefore B is m-dissipative. Finally, for all $u, v \in D(B)$, we have, by (2.1),

$$\langle Bu, v \rangle = \langle u, Bv \rangle.$$

Therefore $G(B) \subset G(B^*)$, and by Corollary 2.4.10, it follows that B is self-adjoint. □

Remark 2.6.3. If Ω has a bounded boundary of class C^2, then $D(B) = H^2(\Omega) \cap H_0^1(\Omega)$, with equivalent norms (see Brezis [2], Theorem IX.25, p. 187, or Friedman [1], Theorem 17.2, p. 67).

2.6.2. The Laplacian in an open subset of $\mathbb{R}^N$: C_0 theory

Let Ω be a bounded open subset of $\mathbb{R}^N$, and let $Z = L^\infty(\Omega)$. We define the linear operator C in Z by

$$\begin{cases} D(C) = \{u \in H_0^1(\Omega) \cap Z, \Delta u \in Z\}, \\ Cu = \Delta u, \quad \forall u \in D(C). \end{cases}$$

Proposition 2.6.4. C is m-dissipative in Z.

Proof. First, let us show that C is dissipative. Let $\lambda > 0$, $f \in Z$, and let $M = \|f\|_{L^\infty}$. Let $u \in H_0^1(\Omega)$ be a solution of

$$u - \lambda \Delta u = f,$$

in $\mathcal{D}'(\Omega)$. In particular, this equation is satisfied in $L^2(\Omega)$, and we have

$$(u - M) - \lambda \Delta (u - M) = f - M,$$

in $L^2(\Omega)$. On the other hand, $v = (u - M)^+ \in H_0^1(\Omega)$, with $\nabla v = 1_{\{|u|>M\}} \nabla u$ (Corollary 1.3.6). Applying Lemma 2.6.2, we obtain

$$\int v^2\,dx + \varphi \int_{\{|u|>M\}} |\nabla u|^2\,dx = \int (f - M) v\,dx \leq 0.$$

Therefore $\int v^2 \, dx \leq 0$, and so $v = 0$. We conclude that $u \leq M$ a.e. on Ω. Similarly, we show that $u \geq -M$ a.e. on Ω. Hence $u \in L^\infty(\Omega)$, and $\|u\|_{L^\infty} \leq \|f\|_{L^\infty}$. It follows that C is dissipative. Now let $f \in L^\infty(\Omega) \subset L^2(\Omega)$. By §2.6.1, there exists $u \in H_0^1(\Omega)$, with $\Delta u \in L^2(\Omega)$, a solution of $u - \Delta u = f$, in $L^2(\Omega)$. We already know that $u \in L^\infty(\Omega)$, so that $u \in D(C)$, and $u - Cu = f$. Therefore C is m-dissipative. □

Lemma 2.6.5. *If Ω has a Lipschitz continuous boundary, then*

$$D(C) \subset C_0(\overline{\Omega}) = \{u \in C(\overline{\Omega});\ u_{|\partial\Omega} = 0\}.$$

Proof. The proof is difficult, and uses the notion of a barrier function (see Gilbarg and Trudinger [1], Theorem 8.30, p. 206). □

Remark 2.6.6. It follows from Lemma 2.6.5 that in general the domain of C is not dense in Z. The fact that the domain is dense will turn out to be very important (see Chapter 3). This is the reason why we are led to consider another example.

We now set $X = C_0(\overline{\Omega})$, and we define the operator A as follows:

$$\begin{cases} D(A) = \{u \in X \cap H_0^1(\Omega), \Delta u \in X\}, \\ Au = \Delta u, \quad \forall u \in D(A). \end{cases}$$

Proposition 2.6.7. *Assume that Ω has a Lipschitz continuous boundary. Then A is m-dissipative, with dense domain.*

Proof. $\mathcal{D}(\Omega)$ is dense in X, and $\mathcal{D}(\Omega) \subset D(A)$; and so $D(A)$ is dense in X. On the other hand, X is equipped with the norm of $L^\infty(\Omega)$, and so $X \hookrightarrow Z$ and $G(A) \subset G(C)$. Since C is dissipative, A is also dissipative. Now let $f \in X \hookrightarrow L^\infty(\Omega)$. Since C is m-dissipative, there exists $u \in D(C)$, such that $u - \Delta u = f$. By Lemma 2.6.5, we have $u \in X$, and so $\Delta u \in X$. Therefore, $u \in D(A)$ and $u - Au = f$. Hence A is m-dissipative. □

Remark 2.6.8. In the three examples of §2.6.1 and §2.6.2, note that the same formula (the Laplacian), corresponds to several operators that enjoy different properties (since they are defined in different domains). In particular, the expression the operator Δ has a meaning only if we specify the space in which this operator applies and its domain.

2.6.3. The wave operator (or the Klein–Gordon operator) in $H_0^1(\Omega) \times L^2(\Omega)$

Let Ω be any open subset of $\mathbb{R}^N$, and let $X = H_0^1(\Omega) \times L^2(\Omega)$. We deal either with real-valued functions or with complex-valued functions, but in both cases X is considered as a real Hilbert space (see §2.5). Let

$$\lambda = \inf\{\|\nabla u\|_{L^2}, u \in H_0^1(\Omega), \|u\|_{L^2} = 1\}. \tag{2.2}$$

(In the case in which Ω is bounded, we recall that λ is the first eigenvalue of $-\Delta$ in $H_0^1(\Omega)$, and that $\lambda > 0$). Let $m > -\lambda$. Then X can be equipped with the scalar product

$$\langle (u,v), (w,z) \rangle = \int (\nabla u \cdot \nabla v + muw + vz)\,dx.$$

This scalar product defines a norm on X which is equivalent to the usual norm. We define the linear operator A in X by

$$\begin{cases} D(A) = \{(u,v) \in X, \Delta u \in L^2(\Omega), v \in H_0^1(\Omega)\}; \\ A(u,v) = (v, \Delta u - mu), \quad \forall (u,v) \in D(A). \end{cases}$$

Proposition 2.6.9. *A is skew-adjoint, and in particular A and $-A$ are m-dissipative with dense domains.*

Proof. $\mathcal{D}(\Omega) \times \mathcal{D}(\Omega) \subset D(A)$ and so $D(A)$ is dense in X. On the other hand, for all $((u,v),(w,z)) \in D(A)^2$, and by (2.1), we have

$$\langle A(u,v), (w,z) \rangle = \int (\nabla v \cdot \nabla w + mvw + (\Delta u - mu)z)\,dx$$
$$= -\int (\nabla u \cdot \nabla z + muz + (\Delta w - mw)v)\,dx$$
$$= -\langle (u,v), A(w,z) \rangle. \tag{2.3}$$

Applying (2.3) with $(u,v) = (w,z)$, it follows that

$$\langle A(u,v), (u,v) \rangle = 0.$$

Hence A is dissipative (Proposition 2.4.2). Now let $(f,g) \in X$. The equation $(u,v) - A(u,v) = (f,g)$ is equivalent to the following system:

$$\begin{cases} 2u - \Delta u = f + g; & (2.4) \\ v = u - f. & (2.5) \end{cases}$$

By Proposition 2.6.1, there exists a solution $u \in H_0^1(\Omega)$ of (2.4), satisfying $\Delta u \in L^2(\Omega)$. Next, we solve (2.5) and we obtain $v \in H_0^1(\Omega)$. Therefore $(u,v) \in D(A)$ and $(u,v) - A(u,v) = (f,g)$, so that A is m-dissipative. Similarly, we show that $-A$ is m-dissipative. By (2.3), we have $G(A) \subset G(-A^*)$. Corollary 2.4.11 proves that A is skew-adjoint. □

2.6.4. The wave operator (or the Klein–Gordon operator) in $L^2(\Omega) \times H^{-1}(\Omega)$

Let Ω and m be as in §2.6.3. We recall that $H_0^1(\Omega) \hookrightarrow L^2(\Omega) \hookrightarrow (H_0^1(\Omega))' = H^{-1}(\Omega)$ with dense embeddings. We equip $H_0^1(\Omega)$ with the scalar product defined in §2.6.3. Theorem 1.1.4 shows that

$$H^{-1}(\Omega) = \{u \in \mathcal{D}'(\Omega), \exists \varphi_u \in H_0^1(\Omega), \Delta \varphi_u - m\varphi_u = u \text{ in } \mathcal{D}'(\Omega)\}, \qquad (2.6)$$

and that we can equip $H^{-1}(\Omega)$ with the scalar product

$$\langle u, v \rangle_{-1} = \int (\nabla \varphi_u \cdot \nabla \varphi_v + m\varphi_u \varphi_v) dx.$$

Set $Y = L^2(\Omega) \times H^{-1}(\Omega)$. We deal either with real-valued functions or with complex-valued functions, but in both cases X is considered as a real Hilbert space (see §2.5). We define the linear operator B in Y by

$$\begin{cases} D(B) = H_0^1(\Omega) \times L^2(\Omega); \\ B(u,v) = (v, \Delta u - mu) \in Y, \quad \forall (u,v) \in D(B). \end{cases}$$

Proposition 2.6.10. *B is skew-adjoint. In particular, B and $-B$ are m-dissipative with dense domains.*

Proof. $\mathcal{D}(\Omega) \times \mathcal{D}(\Omega) \subset D(B)$ and so $D(B)$ is dense in Y. Let $((u,v), (w,z)) \in D(B)^2$, and consider φ_v and φ_z defined by (2.6). Since $v, z \in L^2(\Omega)$, we have $\Delta \varphi_v, \Delta \varphi_z \in L^2(\Omega)$. Applying (2.1), we obtain

$$\begin{aligned}
\langle B(u,v), (w,z) \rangle_{L^2 \times H^{-1}} &= \int vw\, dx + \langle \Delta u - mu, z \rangle_{-1} \\
&= \int vw\, dx + \int (\nabla u \cdot \nabla \varphi_z + mu\varphi_z)\, dx \\
&= \int vw\, dx - \int u(\Delta \varphi_z - m\varphi_z)\, dx \\
&= \int vw\, dx - \int uz\, dx.
\end{aligned}$$

Similarly, we have

$$\langle (u,v), B(w,z) \rangle_{L^2 \times H^{-1}} = \int zu\, dx - \int wv\, dx.$$

Therefore,

$$\langle B(u,v), (w,z) \rangle_{L^2 \times H^{-1}} = -\langle (u,v), B(w,z) \rangle_{L^2 \times H^{-1}}. \qquad (2.7)$$

Applying (2.7) with $(u,v) = (w,z)$, it follows that

$$\langle B(u,v), (u,v) \rangle = 0.$$

Thus, B is dissipative (Proposition 2.4.2). Now let $(f,g) \in Y$. The equation $(u,v) - B(u,v) = (f,g)$ is equivalent to the system (2.4)–(2.5) of §2.6.3. By Theorem 1.1.4 (see the proof of Proposition 2.6.1), there exists a solution $u \in H_0^1(\Omega)$ of (2.4). Next, we solve (2.5) and we obtain $v \in L^2(\Omega)$. Therefore $(u,v) \in D(B)$ and $(u,v) - B(u,v) = (f,g)$; hence B is m-dissipative. Similarly, we show that $-B$ is m-dissipative. By (2.7), we have $G(B) \subset G(-B^*)$. Corollary 2.4.11 proves that B is skew-adjoint. □

Proposition 2.6.11. *We use the same notation as in §2.6.4. Then Y and B are the extensions of X and A given by Proposition 2.3.1.*

Proof. Properties (i), (iii), and (iv) are clearly satisfied. We need only show (ii), i.e.

$$\|U\|_Y \approx \|(I-A)^{-1}U\|_X, \quad \forall U \in X.$$

Let $U \in X$ and $V \in D(A)$ be such that $U = (I-A)V$. We show that $\|(I-A)V\|_Y \approx \|V\|_X$. Indeed, since B is skew-adjoint, we have

$$\|(I-A)V\|_Y^2 = \langle (I-B)V, (I-B)V \rangle_Y = \|V\|_Y^2 + \|BV\|_Y^2.$$

Let $V = (u,v)$. We have

$$\|BV\|_Y^2 = \|v\|_{L^2}^2 + \|\Delta u - mu\|_{H^{-1}}^2 = \|v\|_{L^2}^2 + \|u\|_{H^1}^2 = \|V\|_X^2;$$

hence the result. □

2.6.5. The Schrödinger operator

Let Ω be any open subset of $\mathbb{R}^N$, and let $Y = L^2(\Omega, \mathbb{C})$. Y is considered as a real Hilbert space (see §2.5). We define the linear operator B in Y by

$$\begin{cases} D(B) = \{u \in H_0^1(\Omega, \mathbb{C}), \Delta u \in Y\}, \\ Bu = i\Delta u, \quad \forall u \in D(B). \end{cases}$$

In what follows, we write $L^2(\Omega)$ and $H_0^1(\Omega)$ instead of $L^2(\Omega, \mathbb{C})$ and $H_0^1(\Omega, \mathbb{C})$.

Proposition 2.6.12. *B is skew-adjoint, and in particular B and $-B$ are m-dissipative with dense domains.*

Proof. The result follows from Proposition 2.6.1 and Corollary 2.5.2. □

Remark 2.6.13. As in §2.6.1, if Ω has a bounded boundary of class C^2, then $D(B) = H^2(\Omega) \cap H_0^1(\Omega)$, with equivalent norms.

We now set $X = H^{-1}(\Omega, \mathbb{C})$ and, given $u \in X$, we denote by $\varphi_u \in H_0^1(\Omega, \mathbb{C})$ the solution of $-\Delta\varphi_u + \varphi_u = u$ in X. We equip X with the scalar product

$$\langle u, v \rangle_{-1} = \langle \varphi_u, \varphi_v \rangle_{H^1} = \operatorname{Re} \int_\Omega \left(\nabla\varphi_u \cdot \overline{\nabla\varphi_v} + \varphi_u \overline{\varphi_v} \right) dx,$$

for $u, v \in X$. We define the linear operator C in X by

$$\begin{cases} D(C) = H_0^1(\Omega); \\ Cu = \Delta u, \quad \forall u \in D(C). \end{cases}$$

Proposition 2.6.14. *C is self-adjoint ≤ 0.*

Proof. We have $\mathcal{D}(\Omega, \mathbb{C}) \subset D(C)$ so that $D(C)$ is dense in X. Furthermore, for all $u, v \in D(C)$,

$$\langle Cu, v \rangle_{-1} = \langle Cu - u, v \rangle_{-1} + \langle u, v \rangle_{-1} = \langle u, \varphi_v \rangle_{H^1} + \langle u, v \rangle_{-1}$$
$$= -\langle u, v \rangle_{L^2} + \langle u, v \rangle_{-1}. \tag{2.8}$$

Taking $u = v$, it follows that

$$\langle Cu, u \rangle_{-1} = -\|u\|_{L^2}^2 + \|u\|_{H^{-1}}^2 \leq 0,$$

and so C is dissipative. Theorem 1.1.4 proves that C is m-dissipative. By (2.8), we have

$$\langle Cu, v \rangle_{-1} = \langle u, Cv \rangle_{-1},$$

for all $u, v \in D(C)$. It follows that $G(C) \subset G(C^*)$, and so C is self-adjoint (Corollary 2.4.10). $\square$

Finally, consider the operator A in X given by

$$\begin{cases} D(A) = H_0^1(\Omega); \\ Au = i\Delta u, \quad \forall u \in D(A). \end{cases}$$

Applying Proposition 2.6.14 and Corollary 2.5.2, we obtain the following result.

Corollary 2.6.15. *A is skew-adjoint, and in particular A and $-A$ are m-dissipative with dense domains.*

Notes. For more information about §2.6, see Brezis [2], Courant and Hilbert [1], as well as Gilbarg and Trudinger [1].

3
The Hille–Yosida–Phillips Theorem and applications

3.1. The semigroup generated by an m-dissipative operator

Let X be a Banach space and let A be an m-dissipative operator in X, with dense domain. For $\lambda > 0$, we consider the operators J_λ and A_λ defined in §2.2, and we set $T_\lambda(t) = e^{tA_\lambda}$, for $t \geq 0$.

Theorem 3.1.1. *For all $x \in X$, the sequence $u_\lambda(t) = T_\lambda(t)x$ converges uniformly on bounded intervals of $[0, T]$ to a function $u \in C([0, \infty), X)$, as $\lambda \downarrow 0$. We set $T(t)x = u(t)$, for all $x \in X$ and $t \geq 0$. Then*

$$T(t) \in \mathcal{L}(X) \text{ and } \|T(t)\| \leq 1, \quad \forall t \geq 0; \tag{3.1}$$
$$T(0) = I; \tag{3.2}$$
$$T(t+s) = T(t)T(s), \forall s, t \geq 0. \tag{3.3}$$

In addition, for all $x \in D(A)$, $u(t) = T(t)x$ is the unique solution of the problem

$$\begin{cases} u \in C([0, \infty), D(A)) \cap C^1([0, \infty), X); & (3.4) \\ u'(t) = Au(t), \quad \forall t \geq 0; & (3.5) \\ u(0) = x. & (3.6) \end{cases}$$

Finally,
$$T(t)Ax = AT(t)x. \tag{3.7}$$
for all $x \in D(A)$ and $t \geq 0$.

Proof. We proceed in five steps.

Step 1. By Definition 2.2.11, for all $t \geq 0$ and all $\lambda > 0$, we have

$$T_\lambda(t) = e^{\frac{t}{\lambda}J_\lambda}e^{-\frac{t}{\lambda}I} = e^{-\frac{t}{\lambda}}e^{\frac{t}{\lambda}J_\lambda};$$

and so,

$$\|T_\lambda(t)\| \leq e^{-\frac{t}{\lambda}}e^{\frac{t}{\lambda}\|J_\lambda\|} \leq 1.$$

In particular,
$$\|u_\lambda(t)\| \le \|x\|, \tag{3.8}$$
for all $\lambda > 0$ and all $t \ge 0$.

Step 2. Assume that $x \in D(A)$. It is clear by construction that A_λ and A_μ commute, for all $\lambda, \mu > 0$. In particular, for all $s, t \ge 0$, we have
$$\frac{d}{ds}\{T_\lambda(st)T_\mu(t-st)\} = tT_\lambda(st)T_\mu(t-st)(A_\lambda - A_\mu).$$
It follows that
$$\|u_\lambda(t) - u_\mu(t)\| = \|T_\lambda(t)x - T_\mu(t)x\|$$
$$\le \left\|\int_0^1 \frac{d}{ds}\{T_\lambda(st)T_\mu(t-st)x\}\, ds\right\| \le t\|A_\lambda x - A_\mu x\|.$$
We deduce (Proposition 2.2.12) that u_λ is a Cauchy sequence in $C([0,T], X)$, for all $T > 0$. Let $u \in C([0, \infty), X)$ be its limit.

Step 3. Set $u(t) = T(t)x$. By (3.8), we have
$$\|T(t)x\| \le \|x\|,$$
for all $t \ge 0$, $x \in D(A)$; and so $T(t)$ can be extended to a unique operator $T(t) \in \mathcal{L}(X)$ satisfying $\|T(t)\| \le 1$, for all $t \ge 0$. Take $x \in X$, and $(x_n)_{n\ge 0} \subset D(A)$, such that $x_n \longrightarrow x$ as $n \to \infty$. We have
$$\|T_\lambda(t)x - T(t)x\| \le \|T_\lambda(t)x - T_\lambda(t)x_n\| + \|T_\lambda(t)x_n - T(t)x_n\|$$
$$+ \|T(t)x_n - T(t)x\|$$
$$\le 2\|x_n - x\| + \|T_\lambda(t)x_n - T(t)x_n\|;$$
and so $T_\lambda(t)x \longrightarrow T(t)x$ as $\lambda \downarrow 0$ uniformly on $[0, T]$ for all $T > 0$. Properties (3.1) and (3.2) follow. To show (3.3), it suffices to remark that $T_\lambda(t)T_\lambda(s) = T_\lambda(t+s)$, and so
$$\|T(t)T(s)x - T(t+s)x\| \le \|T(t)T(s)x - T(t)T_\lambda(s)x\|$$
$$+ \|T(t)T_\lambda(s)x - T_\lambda(t)T_\lambda(s)x\|$$
$$+ \|T_\lambda(t+s)x - T(t+s)x\|.$$
It follows that $\|T(t)T(s)x - T(t+s)x\| \longrightarrow 0$ as $\lambda \downarrow 0$.

Step 4. Returning to the case in which $u \in D(A)$, set $v_\lambda(t) = A_\lambda T_\lambda(t)x = T_\lambda(t)A_\lambda x = u'_\lambda(t)$. We have
$$\|v_\lambda(t) - T(t)Ax\| \le \|T(t)Ax - T_\lambda(t)Ax\| + \|A_\lambda x - Ax\|.$$

Hence, $v_\lambda \longrightarrow T(t)Ax$ as $\lambda \downarrow 0$, uniformly on $[0,T]$ for all $T > 0$. Taking

$$u_\lambda(t) = x + \int_0^t v_\lambda(s)\,ds,$$

and letting $\lambda \downarrow 0$, it follows that

$$u(t) = x + \int_0^t T(s)Ax\,dx.$$

Thus $u \in C^1([0,\infty), X)$, and

$$u'(t) = T(t)Ax, \tag{3.9}$$

for all $t \geq 0$. Finally, we have $v_\lambda(t) = A(J_\lambda T_\lambda(t)x)$, and

$$\|J_\lambda T_\lambda(t)x - T(t)x\| \leq \|T_\lambda(t)x - T(t)x\| + \|J_\lambda T(t)x - T(t)x\|.$$

Therefore, $(J_\lambda T_\lambda(t)x, A(J_\lambda T_\lambda(t)x)) \longrightarrow (T(t)x, T(t)Ax)$ in $X \times X$ as $\lambda \downarrow 0$. Since $G(A)$ is closed, it follows that $T(t)x \in D(A)$ for all $t \geq 0$, and $AT(t)x = T(t)Ax$, hence (3.7). We conclude that $u \in C([0,\infty), D(A))$. Putting together (3.7) and (3.9), we obtain (3.5).

Step 5. Uniqueness of the solution of (3.4)–(3.6). Let u be a solution, and let $\tau > 0$. Set

$$v(t) = T(\tau - t)u(t),$$

for $t \in [0,\tau]$. We have $v \in C([0,t], D(A)) \cap C^1([0,t], X)$, and

$$v'(t) = -AT(\tau - t)u(t) + T(\tau - t)u'(t) = T(\tau - t)[u'(t) - Au(t)] = 0,$$

for all $t \in [0,\tau]$. Hence, $v(\tau) = v(0)$, and so $u(\tau) = T(\tau)x$. $\tau \geq 0$ being arbitrary, the proof is complete. $\square$

3.2. Two important special cases

We assume in this section that X is a real Hilbert space. The following result sharpens the conclusions of Theorem 3.1.1.

Theorem 3.2.1. *Assume that A is self-adjoint ≤ 0. Let $x \in X$, and let $u(t) = T(t)x$. Then u is the unique solution of the following problem:*

$$\begin{cases} u \in C([0,\infty), X) \cap C((0,\infty), D(A)) \cap C^1((0,\infty), X); & (3.10) \\ u'(t) = Au(t), \quad \forall t > 0; & (3.11) \\ u(0) = x. & (3.12) \end{cases}$$

In addition, we have

$$\|Au(t)\| \le \frac{1}{t\sqrt{2}}\|x\|; \tag{3.13}$$

$$-\langle Au(t), u(t)\rangle \le \frac{1}{2t}\|x\|^2. \tag{3.14}$$

Finally,

$$\|Au(t)\|^2 \le -\frac{1}{2t}\langle Ax, x\rangle, \tag{3.15}$$

if $x \in D(A)$.

Proof. We easily verify that A_λ is self-adjoint ≤ 0, for all $\lambda > 0$. If $u_\lambda(t) = T_\lambda(t)x$, the functions $\|u_\lambda(t)\|$ and $\|u'_\lambda(t)\|$ are non-increasing with respect to t. In addition, we have

$$\frac{\mathrm{d}}{\mathrm{d}t}\|u_\lambda(t)\|^2 = 2\langle A_\lambda u_\lambda(t), u_\lambda(t)\rangle, \tag{3.16}$$

$$\frac{\mathrm{d}}{\mathrm{d}t}\langle A_\lambda u_\lambda(t), u_\lambda(t)\rangle = 2\langle A_\lambda u_\lambda(t), u'_\lambda(t)\rangle = 2\|u'_\lambda(t)\|. \tag{3.17}$$

From (3.17), it follows that $-\langle A_\lambda u_\lambda(t), u_\lambda(t)\rangle$ is non-increasing with respect to t. Integrating (3.16) between 0 and $t > 0$, it follows that

$$-t\langle A_\lambda u_\lambda(t), u_\lambda(t)\rangle \le -\int_0^t \langle A_\lambda u_\lambda(s), u_\lambda(s)\rangle\,\mathrm{d}s \le \frac{1}{2}\|x\|^2. \tag{3.18}$$

Integrating (3.17), we obtain

$$2t\|u'_\lambda(t)\| \le 2\int_0^t \|u'_\lambda(s)\|\,\mathrm{d}s = -\langle A_\lambda x, x\rangle + \langle A_\lambda u_\lambda(t), u_\lambda(t)\rangle \le -\langle A_\lambda x, x\rangle. \tag{3.19}$$

and

$$t^2\|u'_\lambda(t)\| \le 2\int_0^t s\|u'_\lambda(s)\|\,\mathrm{d}s = 2\int_0^t s\frac{\mathrm{d}}{\mathrm{d}s}\langle A_\lambda u_\lambda(s), u_\lambda(s)\rangle\,\mathrm{d}s$$

$$\le -\int_0^t \langle A_\lambda u_\lambda(s), u_\lambda(s)\rangle.$$

Hence, with (3.18),

$$2t^2\|u'_\lambda(t)\|^2 \le \|x\|^2. \tag{3.20}$$

On the other hand, the vector subspace $G(A)$ is closed in $X \times X$ strong, and so is closed in $X \times X$ weak. Furthermore, $u'_\lambda(t) = A_\lambda u_\lambda(t) = A_\lambda J_\lambda u_\lambda(t)$, and $J_\lambda u_\lambda(t) \longrightarrow u(t)$ as $\lambda \downarrow 0$ (see Step 4 of the proof of Theorem 3.1.1). In addition (by (3.14)), for all $t > 0$, $\|A_\lambda J_\lambda u_\lambda(t)\|$ is bounded as $\lambda \downarrow 0$. Therefore $u(t) \in D(A)$, for all $t > 0$, with

$$Au(t) = \lim_{\lambda \downarrow 0} A_\lambda J_\lambda u_\lambda(t),$$

in X weak. (3.10), (3.11), and (3.12) now follow from Theorem 3.1.1, and (3.13), (3.14), and (3.15) are obtained by passing to the limit in (3.20), (3.18), and (3.19).

It remains to show the uniqueness of u. To do this, take $t > 0$ and $0 < \varepsilon < t$. It follows from (3.10) and Theorem 3.1.1 that $u(t) = T(t - \varepsilon)u(\varepsilon)$, and so

$$\|u(t) - T(t)x\| \leq \|u(\varepsilon) - x\| + \|T(\varepsilon)x - x\| \longrightarrow 0 \quad \text{as } \varepsilon \downarrow 0.$$

Therefore $u(t) = T(t)x$, for all $t \geq 0$, which completes the proof. □

Remark 3.2.2. Theorem 3.2.1 means that $T(t)$ has a smoothing effect on the initial data. Indeed, even if $x \notin D(A)$, we have $T(t)x \in D(A)$, for all $t > 0$. This is in contrast with the isometry groups generated by skew-adjoint operators.

Theorem 3.2.3. *Assume that A is a skew-adjoint operator. Then $(T(t))_{t \geq 0}$ can be extended to a one-parameter group $T(t) : \mathbb{R} \to \mathcal{L}(X)$ such that*

$$T(t)x \in C(\mathbb{R}, X), \quad \forall x \in X; \tag{3.21}$$

$$\|T(t)x\| = \|x\|, \quad \forall x \in X, t \in \mathbb{R}; \tag{3.22}$$

$$T(0) = I; \tag{3.23}$$

$$T(s + t) = T(s)T(t), \quad \forall s, t \in \mathbb{R}. \tag{3.24}$$

In addition, for all $x \in D(A)$, $u(t) = T(t)x$ satisfies $u \in C(\mathbb{R}, D(A)) \cap C^1(\mathbb{R}, X)$ and

$$u'(t) = Au(t), \tag{3.25}$$

for all $t \in \mathbb{R}$.

Proof. We denote by $(T^+(t))_{t \geq 0}$ and $(T^-(t))_{t \geq 0}$ the semigroups corresponding to A and $-A$. We set

$$T(t) = \begin{cases} T^+(t), & \text{if } t \geq 0; \\ T^-(t), & \text{if } t \leq 0. \end{cases}$$

We easily verify (3.21), (3.22), (3.23), and (3.24) for $x \in D(A)$, and then for $x \in X$ by density. Finally,

$$\frac{d^+ u(t)}{dt}(0) = Ax = -\frac{d^+ u(-t)}{dt}(0) = \frac{d^- u}{dt}(0).$$

(3.25) follows from the last identity and Theorem 3.1.1. □

Remark 3.2.4. It is clear that if $x \notin D(A)$, then $u(t) \notin D(A)$ for all $t \in \mathbb{R}$.

Remark 3.2.5. The conclusions of Theorem 3.2.3 may be satisfied without assuming that A is skew-adjoint. Indeed, it suffices (and the proof would the same) that A and $-A$ are m-dissipative (and X may be any Banach space).

Corollary 3.2.6. *Following the notation of Theorem 3.2.3, we have*
$$T(t)^* = T(-t),$$
for all $t \in \mathbb{R}$.

Proof. Let $x, y \in D(A)$. We have
$$\frac{d}{dt}\langle T(t)x, T(t)y\rangle = \langle AT(t)x, T(t)y\rangle + \langle T(t)x, AT(t)y\rangle = 0.$$
Therefore
$$\langle x, y\rangle = \langle T(t)x, T(t)y\rangle,$$
for all $t \in \mathbb{R}$ and all $x, y \in D(A)$. Taking $y = T(-t)z$, we have $\langle x, T(-t)z\rangle = \langle T(t)x, z\rangle$, for all $x, z \in D(A)$; hence the result, by density. □

3.3. Extrapolation and weak solutions

We know (Theorem 3.1.1) that if $x \in D(A)$ then $T(t)x$ is the solution of (3.4)–(3.6). If X is a Hilbert space and A is a self-adjoint operator, then $T(t)x$ is still the solution of (3.10)–(3.12), even for $x \in X$. However, in general, if $x \notin D(A)$, $T(t)x$ is not differentiable in X and then it cannot satisfy (3.11). We will see that the results of §2.3 allow us to identify $T(t)x$. We follow the notation introduced in §2.3, and we denote by $(T(t))_{t\geq 0}$ and $(\overline{T}(t))_{s\geq 0}$ the semigroups corresponding to A and $\overline{A}$. We begin with the following result.

Lemma 3.3.1. *For all $x \in X$ and all $t \geq 0$, we have $T(t)x = \overline{T}(t)x$.*

Proof. The result is clear for $x \in D(A)$. The general case follows from an usual density and continuity argument. □

Corollary 3.3.2. *Let $x \in X$. Then $u(t) = T(t)x$ is the unique solution of*
$$\begin{cases} u \in C([0,\infty), X) \cap C^1([0,\infty), \overline{X}); \\ u'(t) = \overline{A}u(t), \quad \forall t \geq 0; \\ u(0) = x. \end{cases}$$

Proof. We apply Lemma 3.3.1 and Theorem 3.1.1. □

3.4. Contraction semigroups and their generators

Definition 3.4.1. A one-parameter family $(T(t))_{t\geq 0} \subset \mathcal{L}(X)$ is a contraction semigroup in X provided that
  (i) $\|T(t)\| \leq 1$ for all $t \geq 0$;
  (ii) $T(0) = I$;
  (iii) $T(t+s) = T(t)T(s)$ for all $s, t \geq 0$;
  (iv) for all $x \in X$, the function $t \mapsto T(t)x$ belongs to $C([0, \infty), X)$.

Definition 3.4.2. The generator of $(T(t))_{t\geq 0}$ is the linear operator L defined by
$$D(L) = \left\{ x \in X;\ \frac{T(t)x - x}{h}\ \text{has a limit in}\ X\ \text{as}\ h \downarrow 0 \right\},$$
and
$$Lx = \lim_{h \downarrow 0} \frac{T(t)x - x}{h},$$
for all $x \in D(L)$.

The following proposition justifies the introduction of m-dissipative operators in Chapter 2.

Proposition 3.4.3. Let $(T(t))_{t\geq 0}$ be a contraction semigroup in X and let L be its generator. Then L is m-dissipative and $D(L)$ is dense in X.

Proof. We proceed in three steps.

Step 1. L is dissipative. For all $x \in D(L)$, $\lambda > 0$, and $h > 0$, we have
$$\|x - \lambda \frac{T(h)x - x}{h}\| \geq \|\left(1 + \frac{\lambda}{h}\right)x\| - \frac{\lambda}{h}\|T(h)x\| \geq \|x\|;$$
hence the result, letting $h \downarrow 0$.

Step 2. L is m-dissipative. We define the operator J by
$$Jx = \int_0^\infty e^{-t} T(t)x\, dt,$$
for all $x \in X$. It is clear that $J \in \mathcal{L}(X)$, with $\|J\| \leq 1$. For $x \in X$ and $h > 0$, we have
$$\frac{T(h) - I}{h} Jx = \frac{1}{h}\int_0^\infty e^{-t}(T(t+h)x - T(t)x)\, dt$$
$$= \frac{1}{h}\int_h^\infty e^{-(t-h)} T(t)x\, dt - \frac{1}{h}\int_0^\infty e^{-t} T(t)x\, dt$$
$$= \frac{e^h - 1}{h}\int_0^\infty e^{-t} T(t)x\, dt - \frac{e^h}{h}\int_0^h e^{-t} T(t)x\, dt.$$

Letting $h \downarrow 0$, we obtain

$$\lim_{h \downarrow 0} \frac{T(h) - I}{h} Jx = Jx - x;$$

and so $Jx \in D(L)$, with $LJx = Jx - x$, i.e. $Jx - LJx = x$.

Step 3. For all $x \in X$ and $t > 0$, we set

$$x_t = \frac{1}{t} \int_0^t T(s)x \, ds.$$

It is clear that $x_t \longrightarrow x$ as $t \downarrow 0$. To show that $D(L)$ is dense, it suffices to prove that $x_t \in D(L)$, for all $t > 0$. Now we have, for all $h > 0$,

$$t \frac{T(h) - I}{h} x_t = \frac{1}{h} \int_h^{t+h} T(s)x \, ds - \frac{1}{h} \int_0^t T(s)x \, ds$$

$$= \frac{1}{h} \int_t^{t+h} T(s)x \, ds - \frac{1}{h} \int_0^h T(s)x \, ds.$$

As $h \downarrow 0$, the term on the right-hand side converges to $T(t)x - x$, and so $x_t \in D(L)$ with $tLx_t = T(t)x - x$. $\square$

Theorem 3.4.4. (The Hille–Yosida–Phillips Theorem) *A linear operator A is the generator of a contraction semigroup in X if and only if A is m-dissipative with dense domain.*

Proof. If A is the generator of a contraction semigroup in X, Proposition 3.4.3 shows that A is m-dissipative with dense domain.

Conversely, assume that A is m-dissipative with dense domain, and let $(T(t))_{t \geq 0}$ be the semigroup corresponding to A given by Theorem 3.1.1. Then, $(T(t))_{t \geq 0}$ is clearly a contraction semigroup. Denote its generator by L and let us show that $L = A$.

For all $x \in D(A)$ and $h > 0$, we have (Theorem 3.1.1)

$$T(h)x = x + \int_0^t T(s)Ax \, ds,$$

and so $x \in D(L)$ with $Lu = Au$. Consequently, $G(A) \subset G(L)$.

Finally, let $y \in D(L)$. Since A is m-dissipative, there exists $x \in D(A)$ such that $x - Ax = y - Ly$; and since $G(A) \subset G(L)$, we have $(x - y) - L(x - y) = 0$. L being dissipative, we have $x = y$, and so $G(L) \subset G(A)$. It follows that $A = L$, which completes the proof. $\square$

Finally, the following result shows the uniqueness of the semigroup generated by an m-dissipative operator with dense domain.

Proposition 3.4.5. *Let A be an m-dissipative operator with dense domain. Assume that A is the generator of a contraction semigroup $(S(t))_{t\geq 0}$. Then $(S(t))_{t\geq 0}$ is the semigroup corresponding to A given by Theorem 3.1.1.*

Proof. Let $(T(t))_{t\geq 0}$ be the semigroup corresponding to A given by Theorem 3.1.1. Let $x \in D(A)$, and $u(t) = S(t)x$. For all $t \geq 0$ and $h > 0$, we have
$$\frac{u(t+h)-u(t)}{h} = \frac{S(h)-I}{h}u(t) = S(t)\frac{S(h)x-x}{h} \longrightarrow S(t)Ax \quad \text{as } h \downarrow 0.$$
We deduce that $S(t)x \in D(A)$, for all $t \geq 0$, and that
$$AS(t)x = S(t)Ax = \frac{\mathrm{d}^+ u}{\mathrm{d}t}(t),$$
for all $t \geq 0$. Thus $u \in C([0,\infty), D(A)) \cap C^1([0,\infty), X)$ and $u'(t) = Au(t)$, for $t \geq 0$. Therefore, by Theorem 3.1.1, we have $S(t)x = T(t)x$; hence the result, by density. □

The following definition is related to Theorem 3.2.3.

Definition 3.4.6. A one-parameter family $(T(t))_{t\in\mathbb{R}}$ of linear operators is said to be an isometry group in X provided that
(i) $\|T(t)x\| = \|x\|$ for all $x \in X$ and all $t \in \mathbb{R}$;
(ii) $T(0) = I$;
(iii) $T(t+s) = T(t)T(s)$ for all $s, t \in \mathbb{R}$;
(iv) for all $x \in X$ the function $t \mapsto T(t)x$ belongs to $C(\mathbb{R}, X)$.

Further to Theorem 3.2.3 and Remark 3.2.5, we have the following result.

Proposition 3.4.7. *Let A be an m-dissipative operator with dense domain, and let $(T(t))_{t\geq 0}$ be the contraction semigroup generated by A. Then $(T(t))_{t\geq 0}$ is the restriction to $\mathbb{R}_+$ of an isometry group if and only if $-A$ is m-dissipative.*

Proof. It is clear by Theorem 3.2.3 and Remark 3.2.5 that the condition $-A$ is m-dissipative is sufficient. Assume that $(T(t))_{t\geq 0}$ is the restriction to $\mathbb{R}_+$ of an isometry group $(T(t))_{t\in\mathbb{R}}$, and set $U(t) = T(-t)$, for $t \geq 0$. Then $(U(t))_{t\geq 0}$ is a contraction semigroup. Let B be its generator. For all $h > 0$ and $x \in X$, we have
$$\frac{U(h)-I}{h}x = \frac{T(-h)-I}{h}x = -U(h)\frac{T(h)-I}{h}x.$$
We deduce immediately that $B = -A$; hence the result. □

3.5. Examples in the theory of partial differential equations

3.5.1. The heat equation

We use the notation of §2.6.1, and we denote by $(S(t))_{t\geq 0}$ the semigroup generated by B in Y.

Lemma 3.5.1. *The embedding $D(B) \hookrightarrow H_0^1(\Omega)$ is continuous.*

Proof. Let $u \in D(B)$. In particular, we have $u \in H_0^1(\Omega)$ and, by Lemma 2.6.2,
$$\|u\|_{H^1}^2 = \|u\|_{L^2}^2 + \|\nabla u\|_{L^2}^2 = \|u\|_{L^2}^2 - \langle Bu, u\rangle.$$
Thus,
$$\|u\|_{H^1} \leq 2\|u\|_{D(B)}.$$
for all $u \in D(B)$. □

Applying Proposition 2.6.1 and Theorem 3.2.1, we obtain the following proposition.

Proposition 3.5.2. *Let $\varphi \in L^2(\Omega)$ and let $u(t) = S(t)\varphi$ for $t \geq 0$. Then u is the unique solution of the problem*
$$\begin{cases} u \in C([0,\infty), L^2(\Omega)) \cap C^1((0,\infty), L^2(\Omega)), \Delta u \in C((0,\infty), L^2(\Omega)); & (3.26) \\ u'(t) = \Delta u(t), \quad \forall t > 0; & (3.27) \\ u(0) = \varphi. & (3.28) \end{cases}$$
In addition, we have
$$u \in C((0,\infty), H_0^1(\Omega)); \tag{3.29}$$
$$\|\Delta u\|_{L^2} \leq \frac{1}{t\sqrt{2}}\|\varphi\|_{L^2}, \quad \forall t > 0; \tag{3.30}$$
$$\|\nabla u\|_{L^2} \leq \frac{1}{\sqrt{t\sqrt{2}}}\|\varphi\|_{L^2}, \quad \forall t > 0. \tag{3.31}$$

Assuming more regularity on φ, the solution u is also more regular.

Proposition 3.5.3. *In Proposition 3.5.2, assume further that $\varphi \in H_0^1(\Omega)$. Then $u \in C([0,\infty), H_0^1(\Omega))$ and*
$$\|\Delta u\|_{L^2} \leq \frac{1}{\sqrt{t\sqrt{2}}}\|\nabla\varphi\|_{L^2}, \tag{3.32}$$
for all $t > 0$. In addition, if $\Delta\varphi \in L^2(\Omega)$, then $u \in C^1([0,\infty), L^2(\Omega))$, $\Delta u \in C([0,\infty), L^2(\Omega))$ and (3.27) is satisfied for $t = 0$.

Proof. Assume first that $\varphi \in H_0^1(\Omega)$ and $\triangle\varphi \in L^2(\Omega)$. Then the result is a straightforward consequence of Theorem 3.1.1 and (3.15).

By density, (3.32) is verified for all $\varphi \in H_0^1(\Omega)$. Now we need only show that $u \in C([0,\infty), H_0^1(\Omega))$, i.e. (by (3.26)), that $u(t) \to \varphi$ in $H_0^1(\Omega)$, as $t \downarrow 0$. We use the notation introduced in the proof of Theorem 3.2.1. Passing to the limit in (3.19) as $\lambda \downarrow 0$, it follows that

$$\int_0^1 \|\triangle u(s)\|_{L^2}^2 \, ds \leq \frac{1}{2}\|\nabla\varphi\|_{L^2}^2;$$

$$\|\nabla u(t)\|_{L^2}^2 - \|\nabla\varphi\|_{L^2}^2 = 2\int_0^t \|\triangle u(s)\|_{L^2}^2 \, ds.$$

Thus $\|\nabla u(t)\|_{L^2} \longrightarrow \|\nabla\varphi\|_{L^2}$, and $\|u(t)\|_{H^1} \longrightarrow \|\varphi\|_{H^1}$ as $t \downarrow 0$. On the other hand, we know that $u(t) \longrightarrow \varphi$ in L^2 as $t \downarrow 0$; hence the result. □

Remark 3.5.4. If Ω has a bounded boundary of class C^2, then (see Remark 2.6.3) $D(B) = H^2(\Omega) \cap H_0^1(\Omega)$. Therefore, if $\varphi \in H^2(\Omega) \cap H_0^1(\Omega)$, then we have $u \in C([0,\infty), H^2(\Omega))$.

Proposition 3.5.5. *Let λ be defined by (2.2). Then*

$$\|S(t)\|_{\mathcal{L}(L^2)} \leq e^{-\lambda t}, \tag{3.33}$$

for all $t \geq 0$.

Proof. Let $\varphi \in D(B)$, and let $f(t) = (e^{\lambda t}\|S(t)\varphi\|)^2$, for $t \geq 0$. We have

$$e^{-2\lambda t}f'(t) = 2\lambda \int u(t)^2 + 2\int u(t)u'(t)$$

$$= 2\lambda \int u(t)^2 + 2\int u(t)\triangle u(t)$$

$$= 2\lambda \int u(t)^2 - 2\int |\nabla u(t)|^2 \leq 0.$$

Thus $\|S(t)\varphi\| \leq e^{-\lambda t}\|\varphi\|$ for all $t \geq 0$ and all $\varphi \in D(B)$. The general result follows by density. □

We now assume that Ω has a bounded Lipschitz continuous boundary and we follow the notation of §2.6.2. Let $(T(t))_{t\geq 0}$ be the semigroup generated by A in X. We have $X \hookrightarrow Y$, and $G(A) \subset G(B)$. We easily deduce the following result.

Lemma 3.5.6. *For all $\varphi \in X$ and all $t \geq 0$, we have $T(t)\varphi = S(t)\varphi$.*

Consequently, for all $\varphi \in X$, $u(t) = T(t)\varphi$ satisfies the conclusions of Proposition 3.5.2. In addition, the following estimates hold.

Proposition 3.5.7. *Let $1 \leq q \leq p \leq \infty$. Then*

$$\|S(t)\varphi\|_{L^p} \leq (4\pi t)^{-\frac{N}{2}(\frac{1}{p} - \frac{1}{q})} \|\varphi\|_{L^q}, \tag{3.34}$$

for all $t > 0$ and all $\varphi \in X$.

The proof requires the following two results.

Lemma 3.5.8. *For $t > 0$, we define $K(t) \in \mathcal{S}(\mathbb{R}^N)$ by $K(t)x = (4\pi t)^{-\frac{N}{2}} e^{-\frac{|x|^2}{4t}}$. Let $\psi \in C_c(\mathbb{R}^N)$ and let $v(t) = K(t) \star \psi$. Then $v \in C([0, \infty), C_b(\mathbb{R}^N)) \cap C^\infty((0, \infty), C_b^2(\mathbb{R}^N))$ and, for all $1 \leq p \leq \infty$, we have $v \in C([0, \infty), L^p(\mathbb{R}^N)) \cap C^\infty((0, \infty), L^p(\mathbb{R}^N))$. In addition:*

(i) $v_t = \Delta v$ for all $t > 0$;

(ii) $v(0) = \psi$;

(iii) $\|v(t)\|_{L^p} \leq (4\pi t)^{-\frac{N}{2}(\frac{1}{q} - \frac{1}{p})} \|\psi\|_{L^q}$, for $1 \leq q \leq p \leq \infty$ and for all $t > 0$.

Proof. Regularity and properties (i) and (ii) follow from easy calculations. Property (iii) is a consequence of Young's inequality*, since

$$\|K(t)\|_{L^p} = p^{-\frac{N}{2p}} (4\pi t)^{-\frac{N}{2}(1 - \frac{1}{p})} \leq (4\pi t)^{-\frac{N}{2}(1 - \frac{1}{p})},$$

for $1 \leq p \leq \infty$ and for all $t > 0$. □

Lemma 3.5.9. *Let $\varphi \in Y$, $\varphi \geq 0$ a.e. on Ω. Then, for all $t > 0$, we have $S(t)\varphi \geq 0$ a.e. on Ω.*

Proof. By density, we may assume that $\varphi \in D(B)$. We set $u(t) = S(t)\varphi$, and we consider $u^- \in C([0, \infty), H_0^1(\Omega))$. By Proposition 3.5.2, we have, for all $t > 0$,

$$\frac{d}{dt}\int_\Omega (u^-)^2 = -\int_\Omega u_t u^- = -\int_\Omega u^- \Delta u = \int_\Omega \nabla u \cdot \nabla u^- = -\int_\Omega |\nabla u^-|^2 \leq 0.$$

From this, we deduce that $\int_\Omega (u^-)^2 \leq 0$, for all $t > 0$, and so $u \geq 0$. □

---

* Recall Young's inequality: $\|f \star g\|_{L^p} \leq \|f\|_{L^q} \|g\|_{L^r}$, with $1 \leq p, q, r \leq \infty$, and $1/p = 1/q + 1/r - 1$

Proof of Proposition 3.5.7. By density, we may assume that $\varphi \in D(A)$. Let $\zeta = |\varphi|$. Invoking Lemma 3.5.9, for all $t > 0$, we have

$$-S(t)\zeta \leq S(t)\varphi \leq S(t)\zeta,$$

almost everywhere on Ω; and so

$$\|S(t)\varphi\|_{L^p} \leq \|S(t)\zeta\|_{L^p}. \tag{3.35}$$

We define $\psi \in C_c(\mathbb{R}^N)$ by

$$\psi = \begin{cases} \zeta & \text{on } \Omega; \\ 0 & \text{on } \mathbb{R}^N \setminus \Omega. \end{cases}$$

Then we set $v(t) = K(t) \star \psi$, and

$$u(t) = v(t)_{|\Omega} - S(t)\zeta. \tag{3.36}$$

We have $u \in C([0,\infty), C(\overline{\Omega})) \cap C((0,\infty), H^1(\Omega)) \cap C^1((0,\infty), L^2(\Omega))$ and $\Delta u \in C((0,\infty), L^2(\Omega))$. Furthermore, $u(t) = v(t) \geq 0$ on $\partial\Omega$; $u_t = \Delta u$ for $t > 0$; and $u(0) = 0$. Thus,

$$\frac{d}{dt}\int_\Omega (u^-)^2 = -\int_\Omega u_t u^- = -\int_\Omega u^- \Delta u = \int_\Omega \nabla u \cdot \nabla u^- = -\int_\Omega |\nabla u^-|^2 \leq 0,$$

and so $u(t) \leq 0$, for all $t \geq 0$. We then deduce from (3.35) and (3.36) that

$$\|S(t)\varphi\|_{L^p} \leq \|v(t)\|_{L^p}.$$

We conclude by applying Lemma 3.5.8. □

Corollary 3.5.10. *Let $\lambda > 0$ be given by (2.2), and let $M = e^{\lambda|\Omega|^{2/N}/(4\pi)}$. Then*

$$\|S(t)\|_{\mathcal{L}(X)} \leq M e^{-\lambda t}, \tag{3.37}$$

for all $t \geq 0$.

Proof. Let $\varphi \in X$ and let $T > 0$. For $0 \leq t \leq T$, we have

$$\|S(t)\varphi\|_{L^\infty} \leq \|\varphi\|_{L^\infty} \leq e^{-\lambda t}e^{\lambda T}\|\varphi\|_{L^\infty}.$$

For $T \leq t$, it follows from (3.34) that

$$\|S(t)\varphi\|_{L^\infty} \leq (4\pi t)^{-\frac{N}{4}}\|S(t-T)\varphi\|_{L^2} \leq (4\pi t)^{-\frac{N}{4}}e^{-\lambda t}e^{\lambda T}\|\varphi\|_{L^2}$$
$$\leq |\Omega|^{\frac{1}{2}}(4\pi t)^{-\frac{N}{4}}e^{-\lambda t}e^{\lambda T}\|\varphi\|_{L^\infty}.$$

The result follows by taking $T = |\Omega|^{2/N}/(4\pi)$. □

Remark. We cannot take $M = 1$ in (3.37). More precisely, we have

$$\frac{d}{dt}\{\|S(t)\|_{\mathcal{L}(X)}\}_{t=0} = 0;$$

and so, if $\|S(t)\|_{\mathcal{L}(X)} \leq M'e^{-\mu t}$ with $\mu > 0$, we have $M' > 1$. Indeed, let $\varphi \in \mathcal{D}(\Omega)$ be such that $\varphi \equiv 1$ in a neighbourhood of $x_0 \in \Omega$ and $\|\varphi\|_X = 1$, and let $u = S(t)\varphi$. We see that $u \in C^\infty([0, \infty) \times \Omega)$. Thus we have $u_t(0) \equiv 1$ in a neighbourhood of x_0. Consequently, for all $\varepsilon > 0$ and for x in a neighbourhood of x_0, we have $u(t, x) \geq 1 - \varepsilon t$, for t small enough, and so in particular $\|u(t)\|_X \geq 1 - \varepsilon t$, for t small enough; hence the result.

Concerning L^p inequalities, note that applying (3.37) and (3.34), we verify easily that for all $1 \leq p \leq \infty$, there exists a constant M_p such that

$$\|S(t)\|_{\mathcal{L}(L^p)} \leq M_p e^{-\lambda t}, \quad (3.37')$$

for all $t > 0$. Once more, we cannot take $M_p = 1$. Actually, one can see that, for $p \geq 2$, one has

$$\|S(t)\|_{\mathcal{L}(L^p)} \leq e^{-\frac{4\lambda t}{p^2}},$$

for all $t \geq 0$, and that this inequality is optimum in the following sense:

$$\frac{d}{dt}\{\|S(t)\|_{\mathcal{L}(L^p)}\}_{t=0} = -\frac{4\lambda}{p^2}.$$

Indeed, for all $\varphi \in \mathcal{D}(\Omega)$, letting $u(t) = S(t)\varphi$, and multiplying by $|u|^{p-2}u$, we obtain the equation satisfied by u. Applying (3.33), we obtain

$$0 = \frac{1}{p}\frac{d}{dt}\int_\Omega |u(t)|^p + \frac{4}{p^2}\int_\Omega |\nabla u^{\frac{p}{2}}|^2 \geq \frac{1}{p}\frac{d}{dt}\int_\Omega |u(t)|^p + \frac{4\lambda}{p^2}\int_\Omega |u(t)|^p;$$

the inequality follows. To show optimality, it suffices to verify that, for all $\varepsilon > 0$, there exists $\psi \in \mathcal{D}(\Omega)$ such that

$$\int_\Omega |\nabla \psi^{\frac{p}{2}}|^p \leq (\lambda + \varepsilon) \int_\Omega |\psi|^p.$$

To see this, we consider the first eigenfunction φ_1 of $-\Delta$ in $H_0^1(\Omega)$ (see Brezis [2]). For $\varepsilon > 0$, let $\theta_\varepsilon : [0, \infty) \to [0, \infty)$ be such that $\theta_\varepsilon \equiv 0$ in a neighbourhood of 0, $\theta'_\varepsilon(x) \leq 1$ and $x - \theta_\varepsilon(x) \leq \varepsilon$ for all $x \geq 0$. Set $\psi_\varepsilon = (\theta_\varepsilon(\varphi_1))^{2/p}$. We verify that $\psi_\varepsilon \in \mathcal{D}(\Omega)$. Furthermore, $|\nabla \psi_\varepsilon^{p/2}| \leq |\nabla \varphi_1|$ and, last,

$$\int_\Omega \psi_\varepsilon^p \longrightarrow \int_\Omega \varphi_1^2 \quad \text{as } \varepsilon \downarrow 0.$$

Consequently,

$$\int_\Omega |\nabla \psi^{\frac{p}{2}}|^2 \leq \int_\Omega |\nabla \varphi_1|^2 = \lambda \int_\Omega \varphi_1^2 = \lambda \lim_{\varepsilon \downarrow 0} \int_\Omega |\psi_\varepsilon|^p.$$

3.5.2. The wave equation (or the Klein–Gordon equation)

We use the notation introduced in §2.6.3 and §2.6.4, and we denote by $(T(t))_{t \in \mathbb{R}}$ the isometry group generated by A in X, and by $(S(t))_{t \in \mathbb{R}}$ the isometry group generated by B in Y.

Proposition 3.5.11. Let $(\varphi, \psi) \in X$ and let $u(t)$ be the first component of $T(t)(\varphi, \psi)$. Then u is the unique solution of the following problem:

$$\begin{cases} u \in C(\mathbb{R}, H_0^1(\Omega)) \cap C^1(\mathbb{R}, L^2(\Omega)) \cap C^2(\mathbb{R}, H^{-1}(\Omega)); & (3.38) \\ u_{tt} - \Delta u + mu = 0, \quad \text{for all } t \in \mathbb{R}; & (3.39) \\ u(0) = \varphi, \quad u_t(0) = \psi. & (3.40) \end{cases}$$

In addition,

$$\int_\Omega \{|\nabla u(t)|^2 + m|u(t)|^2 + u_t(t)^2\} = \int_\Omega \{|\nabla \varphi|^2 + m|\varphi|^2 + \psi^2\}, \quad (3.41)$$

for all $t \in \mathbb{R}$. Finally, if $(\varphi, \psi) \in D(A)$, we have $u \in C^1(\mathbb{R}, H_0^1(\Omega)) \cap C^2(\mathbb{R}, L^2(\Omega))$ and $\Delta u \in C(\mathbb{R}, L^2(\Omega))$.

Proof. Let $u \in \mathcal{D}'(\mathbb{R}, H^{-1}(\Omega))$ and set $U = (u, u_t)$. Then $U \in C(\mathbb{R}, D(B)) \cap C^1(\mathbb{R}, Y)$ if and only if $u \in C(\mathbb{R}, H_0^1(\Omega)) \cap C^1(\mathbb{R}, L^2(\Omega)) \cap C^2(\mathbb{R}, H^{-1}(\Omega))$. Furthermore, in that case, (3.39)–(3.40) is equivalent to the equation

$$U'(t) = BU(t),$$

for all $t \in \mathbb{R}$. The result then follows from Propositions 2.6.9, 2.6.10, and 2.6.11, Theorem 3.2.3, and Corollary 3.3.2. Note that (3.41) is equivalent to (3.22). □

3.5.3. The Schrödinger equation

We use the notation introduced in §2.6.5, and we denote by $(S(t))_{t \in \mathbb{R}}$ and $(T(t))_{t \in \mathbb{R}}$ the isometry groups generated by B and A. We have $G(B) \subset G(A)$, and from this it is easy to deduce the following result.

Lemma 3.5.12. For all $\varphi \in Y$, we have $S(t)\varphi = T(t)\varphi$, for all $t \in \mathbb{R}$.

Then we have the following.

Proposition 3.5.13. Let $\varphi \in H_0^1(\Omega)$ and let $u(t) = T(t)\varphi$. Then u is the unique solution of the problem

$$\begin{cases} u \in C(\mathbb{R}, H_0^1(\Omega)) \cap C^1(\mathbb{R}, H^{-1}(\Omega)); & (3.42) \\ iu_t + \triangle u = 0, \text{ for all } t \in \mathbb{R}; & (3.43) \\ u(0) = \varphi. & (3.44) \end{cases}$$

In addition,

$$\int_\Omega |u(t)|^2 = \int_\Omega |\varphi|^2, \quad \text{for all } t \in \mathbb{R}, \tag{3.45}$$

$$\int_\Omega |\nabla u(t)|^2 = \int |\nabla \varphi|^2, \quad \text{for all } t \in \mathbb{R}. \tag{3.46}$$

Finally, if $\triangle \varphi \in L^2(\Omega)$, then $u \in C^1(\mathbb{R}, L^2(\Omega))$ and $\triangle u \in C(\mathbb{R}, L^2(\Omega))$.

Proof. We use Theorem 3.2.3, Corollary 3.3.2, and Lemma 3.5.12. (3.45) is equivalent to (3.22). On the other hand, invoking (3.7) and (3.22), we obtain

$$\|Au(t)\|_X = \|A\varphi\|_X, \tag{3.47}$$

for all $t \in \mathbb{R}$. But, for all $v \in H_0^1(\Omega)$, we have

$$\|Av\|_X^2 = \|(\triangle v - v) + v\|_X^2 = \int |\nabla v|^2 - \int |v|^2 + \|v\|_X^2. \tag{3.48}$$

By (3.22), $\|u(t)\|_X = \|\varphi\|_X$ for all $t \in \mathbb{R}$; and so (3.46) follows by putting together (3.47), (3.48), and (3.45). □

3.5.4. The Schrödinger equation in $\mathbb{R}^N$

We use the notation of §3.5.3, and we assume that $\Omega = \mathbb{R}^N$. We can state estimates in the spirit of (3.34).

Proposition 3.5.14. For all $p \in [2, \infty]$ and $t \neq 0$. Then $T(t)$ can be extended to an operator belonging to $\mathcal{L}(L^{p'}(\mathbb{R}^N), L^p(\mathbb{R}^N))$. In addition, we have

$$\|T(t)\|_{\mathcal{L}(L^{p'}, L^p)} \leq \frac{1}{(4\pi|t|)^{N(\frac{1}{2} - \frac{1}{p})}}, \tag{3.49}$$

for all $t \neq 0$.

Proof. Let $\varphi \in \mathcal{S}(\mathbb{R}^N)$ and let $u(t) \in C^\infty(\mathbb{R}, \mathcal{S}(\mathbb{R}^N))$ be defined by

$$\mathcal{F}u(t)(\xi) = e^{-i|\xi|^2 t} \mathcal{F}\varphi(\xi), \tag{3.50}$$

for all $\xi \in \mathbb{R}^N$ and $t \in \mathbb{R}$. We have

$$i\frac{d}{dt}\mathcal{F}u(t)(\xi) - |\xi|^2 \mathcal{F}u(t)(\xi) = 0 \quad \text{in } \mathbb{R}^N,$$

for all $t \in \mathbb{R}$; and so

$$iu_t + \Delta u = 0 \quad \text{in } \mathbb{R}^N,$$

for all $t \in \mathbb{R}$. Therefore we have (Proposition 3.5.13) $u(t) = T(t)\varphi$, for all $t \in \mathbb{R}$. Now we know that

$$\mathcal{F}^{-1}\{e^{-i|\xi|^2 t}\}(x) = \frac{1}{(4\pi t)^{\frac{N}{2}}}e^{\frac{i|x|^2}{4t}} := K(t)x,$$

for all $x \in \mathbb{R}^N$ and $t \neq 0$. It follows from (3.50) that $u(t) = K(t) \star \varphi$ for all $t \neq 0$. We deduce that

$$\|T(t)\varphi\|_{L^\infty} \leq \frac{1}{(4\pi t)^{\frac{N}{2}}}\|\varphi\|_{L^1},$$

for all $t \neq 0$ and $\varphi \in \mathcal{S}(\mathbb{R}^N)$. Thus, one can extend $T(t)$ to an operator of $\mathcal{L}(L^1(\mathbb{R}^N), L^\infty(\mathbb{R}^N))$ such that $\|T(t)\|_{\mathcal{L}(L^1,L^\infty)} \leq (4\pi|t|)^{-\frac{N}{2}}$. Furthermore, $T(t) \in \mathcal{L}(L^2(\mathbb{R}^N), L^2(\mathbb{R}^N))$, with $\|T(t)\|_{\mathcal{L}(L^2,L^2)} = 1$. The general case follows from the Riesz interpolation Theorem (see Dunford and Schwartz [1], p. 525, or Bergh and Löfström [1], p. 2, Theorem 1.1.1). □

Notes. Theorem 3.2.1 can be generalized to the case of the generators of analytic semigroups; see Goldstein [1], Haraux [3], Pazy [1]. One can build semigroups for some classes of operators, m-dissipative operators (non-linear), and maximal monotone operators. These two classes coincide in Hilbert spaces. See Brezis [1], Crandall and Liggett [1], Crandall and Pazy [1], and Haraux [1, 2].

4
Inhomogeneous equations and abstract semilinear problems

Throughout this chapter, we assume that X is a Banach space and that A is an m-dissipative operator with dense domain. We denote by $(T(t))_{t\geq 0}$ the contraction semigroup generated by A.

4.1. Inhomogeneous equations

Let $T > 0$. Given $x \in X$ and $f : [0, T] \to X$, our aim is to solve the problem

$$\begin{cases} u \in C([0,T], D(A)) \cap C^1([0,T], X); & (4.1) \\ u'(t) = Au(t) + f(t), \quad \forall t \in [0,T]; & (4.2) \\ u(0) = x. & (4.3) \end{cases}$$

As in the case of ordinary differential equations, we have the following result (the variation of parameters formula, or Duhamel's formula).

Lemma 4.1.1. *Let $x \in D(A)$ and let $f \in C([0,T], X)$. We consider a solution $u \in C([0,T], D(A)) \cap C^1([0,T], X)$ of problem (4.1)–(4.3). Then, we have*

$$u(t) = T(t)x + \int_0^t T(t-s)f(s)\,ds, \qquad (4.4)$$

for all $t \in [0,T]$.

Proof. Let $t \in (0, T]$. Set

$$w(s) = T(t-s)u(s),$$

for $s \in [0, t]$. Let $s \in [0, t]$ and $h \in (0, t-s]$. We have

$$\frac{w(s+h) - w(s)}{h} = T(t-s-h)\left\{\frac{u(s+h) - u(s)}{h} - \frac{T(h) - I}{h}u(s)\right\}$$
$$\longrightarrow T(t-s)\{u'(s) - Au(s)\} = T(t-s)f(s)$$

as $h \downarrow 0$. Since $T(t-\cdot)f(\cdot) \in C([0,t], X)$, we deduce that $w \in C^1([0,t), X)$ and that

$$w'(s) = T(t-s)f(s), \qquad (4.5)$$

for all $s \in [0, t)$. Integrating (4.5) between 0 and $\tau < t$, and letting $\tau \uparrow t$, we obtain (4.4). □

Corollary 4.1.2. *For all $x \in D(A)$ and $f \in C([0,T], X)$, problem (4.1)–(4.3) has at most one solution.*

Remark 4.1.3. For all $x \in X$ and all $f \in C([0,T], X)$, formula (4.4) defines a function $u \in C([0,T], X)$. Now we are looking for sufficient conditions for u given by (4.4) to be the solution of (4.1)–(4.3).

Remark 4.1.4. It is clear that if u is a solution of (4.1)–(4.3), then $x \in D(A)$. However, this condition is not sufficient. Indeed, assume that $(T(t))_{t \in \mathbb{R}}$ is an isometry group, and let $y \in X \setminus D(A)$. Then (see Remark 3.2.4), $T(t)y \notin D(A)$, for all $t \in \mathbb{R}$. Take $f(t) = T(t)y$, and $x = 0 \in D(A)$. It follows easily that (4.4) gives $u(t) = tT(t)y \notin D(A)$, for $t \neq 0$.

Lemma 4.1.5. *For all $x \in X$ and $f \in L^1((0,T), X)$, formula (4.4) defines a function $u \in C([0,T], X)$. In addition, we have*

$$\|u\|_{C([0,T],X)} \leq \|x\| + \|f\|_{L^1(0,T,X)}.$$

Proof. The result is clear if $f \in C([0,T], X)$, and follows by density in the general case. □

Proposition 4.1.6. *Let $x \in D(A)$ and let $f \in C([0,T], X)$. Assume that at least one of the following conditions is satisfied:*

(i) $f \in L^1((0,T), D(A))$;

(ii) $f \in W^{1,1}((0,T), X)$.

Then u given by (4.4) is the solution of (4.1)–(4.3).

Proof. We proceed in four steps. Set

$$v(t) = \int_0^t T(t-s)f(s)\,ds = \int_0^t T(s)f(t-s)\,ds,$$

for $t \in [0,T]$.

Step 1. We have $v \in C^1([0,T), X)$. Indeed, if $f \in L^1((0,T), D(A))$, for $t \in [0,t)$ and $h \in [0, t-s]$, write

$$\frac{v(t+h) - v(t)}{h} = \int_0^t T(t-s)\frac{T(h) - I}{h}f(s)\,ds + \frac{1}{h}\int_t^{t+h} T(t+h-s)f(s)\,ds,$$

and let $h \downarrow 0$. Note that
$$\frac{T(h) - I}{h} f \longrightarrow Af$$
in $L^1((0, T), X)$, as $h \downarrow 0$, and apply Lemma 4.1.5. It follows that
$$\frac{\mathrm{d}^+ v}{\mathrm{d}t}(t) = \int_0^t T(t - s) Af(s) \, \mathrm{d}s + f(t),$$
for all $t \in [0, T)$. If $f \in W^{1,1}((0, T), X)$, for $t \in [0, T)$ and $h \in [0, T - t]$, we write
$$\frac{v(t + h) - v(t)}{h} = \int_0^t T(s) \frac{f(t + h - s) - f(t - s)}{h} \, \mathrm{d}s + \frac{T(h)}{h} \int_0^h T(t-s) f(s) \, \mathrm{d}s,$$
and we let $h \downarrow 0$. Note that (Corollary 1.4.39)
$$\frac{f(t + h - \cdot) - f(t - \cdot)}{h} \longrightarrow f'(t - \cdot), \text{ as } h \downarrow 0$$
in $L^1((0, t), X)$ and apply Lemma 4.1.5. It follows that
$$\frac{\mathrm{d}^+ v}{\mathrm{d}t}(t) = \int_0^t T(s) f'(t - s) \, \mathrm{d}s + T(t) f(0),$$
for all $t \in [0, T)$. In both cases, we have $\mathrm{d}^+ v/\mathrm{d}t \in C([0, T), X)$; and so $v \in C^1([0, T), X)$.

Step 2. Similarly, we show that $(\mathrm{d}^- v/\mathrm{d}t)(T)$ makes sense and is equal to $\lim_{t \uparrow T} v'(t)$; and so $v \in C^1([0, T], X)$.

Step 3. Let $t \in [0, T)$ and let $h \in [0, T - t]$. We have
$$\frac{T(h) - I}{h} v(t) = \frac{1}{h} \int_0^t T(t + h - s) f(s) \, \mathrm{d}s - \frac{1}{h} \int_0^t T(t - s) f(s) \, \mathrm{d}s$$
$$= \frac{v(t + h) - v(t)}{h} - \frac{1}{h} \int_t^{t+h} T(t + h - s) f(s) \, \mathrm{d}s.$$

Letting $h \downarrow 0$, we deduce $v(t) \in D(A)$, and $Av(t) = v'(t) - f(t)$. This is still true for $t = T$, since the graph of A is closed. It follows that $v \in C([0, T], D(A))$, and that v satisfies (4.2).

Step 4. We have $u(t) = T(t)x + v(t) \in C([0, T], D(A)) \cap C^1([0, T], X)$, and (4.1) follows. Furthermore,
$$u'(t) = AT(t)x + Av(t) + f(t) = Au(t) + f(t),$$
for all $t \in [0, T]$. Hence, we have (4.2), and (4.3) is immediate. □

Corollary 4.1.7. Let $x \in X$, $f \in C([0,T], X)$ and let u be given by (4.4). Then using the notation of §2.3 and §3.3, u is the unique solution of the problem

$$\begin{cases} u \in C([0,T], X) \cap C^1([0,T], \overline{X}); \\ u'(t) = \overline{A}u(t) + f(t), \quad \forall t \in [0,T]; \\ u(0) = x. \end{cases}$$

Proof. We apply Lemma 3.3.1 and Proposition 4.1.6. □

Corollary 4.1.8. Let $x \in X$, $f \in C([0,T], X)$ and let u be given by (4.4). Assume that at least one of the following conditions is satisfied:

(i) $u \in C([0,T], D(A))$;

(ii) $u \in C^1([0,T], X)$.

Then u is the solution of (4.1)–(4.3).

Proof. Assume that (i) holds. By Corollary 4.1.7, we have $u' \in C([0,T], X)$; and so $u \in C^1([0,T], X)$, hence the result.

Now assume that (ii) holds. By Corollary 4.1.7, we have $\overline{A}u \in C([0,T], X)$; and so (Corollary 2.3.2) $u \in C([0,T], D(A))$; hence the result. □

Throughout §4.1, we have supposed that $f \in C([0,T], X)$. But in order to give a sense to (4.4), it suffices that $f \in L^1((0,T), X)$ (Lemma 4.1.5). In this case, we have the following result.

Proposition 4.1.9. Let $x \in X$, $f \in L^1((0,T), X)$ and let $u \in L^1((0,T), X)$. Assume further that $u \in L^1((0,T), D(A))$ or that $u \in W^{1,1}((0,T), X)$. Then u verifies (4.4) if and only if u

$$\begin{cases} u \in L^1((0,T), D(A)) \cap W^{1,1}((0,T), X); \\ u'(t) = Au(t) + f(t), \quad \text{for almost every } t \in [0,T]; \\ u(0) = x. \end{cases}$$

Proof. First note that if $u \in W^{1,1}((0,T), X)$ then $u \in C([0,T], X)$ (Corollary 1.4.36) and so the condition $u(0) = x$ makes sense. Let us first show that the assumptions of the theorem are sufficient to have (4.4). To see this, we argue as in Lemma 4.1.1. We consider $t \in (0,T]$ and we set $w(s) = T(t-s)u(s)$, for almost every $s \in (0,t)$. For all $h \in (0,t)$, and for almost every $s \in (0, t-h)$, we have

$$\frac{w(s+h) - w(s)}{h} = T(t-s-h)\left\{\frac{u(s+h) - u(s)}{h} - \frac{T(h) - I}{h}u(s)\right\}.$$

It follows that w is absolutely continuous from $[0,T]$ to Y. Moreover, the right-hand member converges for almost every $s \in (0,t)$ to $T(t-s)u'(s) - \overline{A}u(s) = T(t-s)f(s)$, as $h \downarrow 0$ (Theorem 1.4.35). Therefore w is right differentiable almost everywhere on $(0,t)$ and

$$\frac{d^+w}{ds}(s) = T(t-s)f(s).$$

Similarly, we show that w is left differentiable almost everywhere on $(0,t)$ and that

$$\frac{d^-w}{ds}(s) = T(t-s)f(s).$$

Consequently (Theorem 1.4.35), $w \in W^{1,1}((0,T),Y)$ and $w'(s) = T(t-s)f(s)$ almost everywhere. We deduce (4.4).

Conversely, assume that u satisfies (4.4). Let $(f_n)_{n \geq 0}$ be a sequence of $C([0,T],X)$ such that $f_n \longrightarrow f$ in $L^1((0,T),X)$ as $n \to \infty$, and let u_n be the corresponding solutions of (4.4). Invoking Corollary 4.1.7, we have

$$u_n'(t) = \overline{A}u_n(t) + f_n(t),$$

for all $t \in [0,T]$; and so

$$u_n(t) = x + \int_0^t (\overline{A}u_n(s) + f_n(s))\,ds,$$

for all $t \in [0,T]$. Letting $n \to \infty$, we obtain (Lemma 4.1.5)

$$u(t) = x + \int_0^t (\overline{A}u(s) + f(s))\,ds,$$

for all $t \in [0,T]$. It follows that $u \in W^{1,1}((0,T),Y)$ and that

$$u'(t) = \overline{A}u(t) + f(t),$$

for almost every $t \in [0,T]$. If $u \in W^{1,1}((0,T),X)$, we have $\overline{A}u \in L^1((0,T),X)$; and so $u \in L^1((0,T),D(A))$ (Corollary 2.3.2). If $u \in L^1((0,T),D(A))$, we have $u' \in L^1((0,T),X)$; and so $u \in W^{1,1}((0,T),X)$. This completes the proof. □

4.2. Gronwall's lemma

In this section, we give a result which is essential in the study of semilinear problems; not only for showing uniqueness of solutions but also for finding bounds on the solutions.

Lemma 4.2.1. (Gronwall's lemma) Let $T > 0$, $\lambda \in L^1(0,T)$, $\lambda \geq 0$ a.e. and $C_1, C_2 \geq 0$. Let $\varphi \in L^1(0,T)$, $\varphi \geq 0$ a.e., be such that $\lambda\varphi \in L^1(0,T)$ and

$$\varphi(t) \leq C_1 + C_2 \int_0^t \lambda(s)\varphi(s)\,\mathrm{d}s,$$

for almost every $t \in (0,T)$. Then we have

$$\varphi(t) \leq C_1 \exp\left(C_2 \int_0^t \lambda(s)\,\mathrm{d}s\right),$$

for almost every $t \in (0,T)$.

Proof. We set

$$\psi(t) = C_1 + C_2 \int_0^t \lambda(s)\varphi(s)\,\mathrm{d}s.$$

ψ is differentiable almost everywhere (since it is absolutely continuous), and we have

$$\psi'(t) \leq C_2\lambda(t)\varphi(t) \leq C_2\lambda(t)\psi(t),$$

for almost every $t \in (0,T)$. Consequently,

$$\frac{\mathrm{d}}{\mathrm{d}t}\left\{\psi(t)\exp\left(C_2 \int_0^t \lambda(s)\,\mathrm{d}s\right)\right\} \leq 0,$$

and so

$$\psi(t) \leq C_1 \exp\left(C_2 \int_0^t \lambda(s)\,\mathrm{d}s\right);$$

hence the result, since $\varphi \leq \psi$. $\square$

Remark 4.2.2. In particular, if $C_1 = 0$, we have $\varphi = 0$ a.e.

4.3. Semilinear problems

Definition 4.3.1. A function $F : X \to X$ is Lipschitz continuous on bounded subsets of X provided that for all $M > 0$, there exists a constant $L(M)$ such that

$$\|F(y) - F(x)\| \leq L(M)\|y - x\|, \quad \forall x, y \in B_M,$$

where B_M is the ball of center 0 and of radius M.

Throughout §4.3, $F : X \to X$ is a Lipschitz continuous function on bounded subsets of X. We denote by $L(M)$ the Lipschitz constant of F in B_M for $M > 0$. In particular, $L(M)$ is a non-decreasing function of M.

Given $x \in X$, we look for $T > 0$ and a solution u of the following problem:

$$\begin{cases} u \in C([0,T], D(A)) \cap C^1([0,T], X); & (4.6) \\ u'(t) = Au(t) + F(u(t)), \quad \forall t \in [0,T]; & (4.7) \\ u(0) = x. & (4.8) \end{cases}$$

We also consider a weak form of the preceding problem. Indeed, by Lemma 4.1.1, any solution u of (4.6)–(4.8) is also solution of the following problem:

$$u(t) = T(t)x + \int_0^t T(t-s)F(u(s))\,ds, \quad \forall t \in [0,T]. \quad (4.9)$$

Finally, note that, for all $u \in C([0,T], X)$, (4.9) is equivalent (following the notation of Corollary 4.1.7) to the problem

$$\begin{cases} u \in C([0,T], X) \cap C^1([0,T], \overline{X}); \\ u'(t) = \overline{A}u(t) + F(u(t)), \quad \forall t \in [0,T]; \\ u(0) = x. \end{cases}$$

4.3.1. A result of local existence

We begin with a uniqueness result.

Lemma 4.3.2. Let $T > 0$, $x \in X$, and let $u, v \in C([0,T], X)$ be two solutions to problem (4.9). Then $u = v$.

Proof. We set $M = \sup\limits_{t \in [0,T]} \max\{\|u(t)\|, \|v(t)\|\}$. We have

$$\|u(t) - v(t)\| \le \int_0^t \|F(u(s)) - F(v(s))\|\,ds \le L(M) \int_0^t \|u(s) - v(s)\|\,ds,$$

and we conclude using Remark 4.2.2. □

Set

$$T_M = \frac{1}{2L(2M + \|F(0)\|) + 2} > 0,$$

for $M > 0$. We can state a first result of local existence.

Proposition 4.3.3. Let $M > 0$ and let $x \in X$ be such that $\|x\| \le M$. Then there exists a unique solution $u \in C([0, T_M], X)$ of (4.9) with $T = T_M$.

Proof. Lemma 4.3.2 proves uniqueness. Let $x \in X$ and let $M \geq \|x\|$. We let $K = 2M + \|F(0)\|$ and

$$E = \{u \in C([0, T_M], X); \|u(t)\| \leq K, \forall t \in [0, T_M]\},$$

and we equip E with the distance generated by the norm of $C([0, T_M], X)$, Let

$$d(u, v) = \max_{t \in [0, T_M]} \|u(t) - v(t)\|,$$

for $u, v \in E$. Since $C([0, T_M], X)$ is a Banach space, (E, d) is a complete metric space. For all $u \in E$, we define $\Phi_u \in C([0, T_M], X)$ by

$$\Phi_u(t) = \mathcal{T}(t)x + \int_0^t \mathcal{T}(t-s)F(u(s))\,ds,$$

for all $t \in [0, T_M]$. Note that for $s \in [0, T_M]$, we have $F(u(s)) = F(0) + (F(u(s)) - F(0))$; and so

$$\|F(u(s))\| \leq \|F(0)\| + KL(K) \leq \frac{M + \|F(0)\|}{T_M}.$$

It follows that

$$\|\Phi_u(t)\| \leq \|x\| + \int_0^t \|F(u(s))\|\,ds \leq M + t\frac{M + \|F(0)\|}{T_M} \leq K.$$

Consequently, we have $F : E \to E$. Furthermore, for all $u, v \in E$, we have

$$\|\Phi_v(t) - \Phi_u(t)\| \leq L(K)\int_0^t \|v(s) - u(s)\|\,ds \leq T_M L(K) d(u, v) \leq \frac{1}{2}d(u, v).$$

Therefore, Φ is a contraction in E with Lipschitz constant $1/2$, and so Φ has a fixed point (Theorem 1.1.1) $u \in E$, which satisfies the requirements of Proposition 4.3.3. □

Theorem 4.3.4. *There exists a function $T : X \to (0, \infty]$ with the following properties: for all $x \in X$, there exists $u \in C([0, T(x)), X)$ such that for all $0 < T < T(x)$, u is the unique solution of (4.9) in $C([0, T], X)$. In addition,*

$$2L(\|F(0)\| + 2\|u(t)\|) \geq \frac{1}{T(x) - t} - 2, \tag{4.10}$$

for all $t \in [0, T(x))$. In particular, we have the following alternatives:

(i) $T(x) = \infty$;

(ii) $T(x) < \infty$ and $\lim_{t \uparrow T(x)} \|u(t)\| = \infty$.

Remark 4.3.5. If property (i) holds, we say that the solution u is global. On the other hand, if (ii) holds, we say that u blows up in finite time. In other words, the alternatives (i)-(ii) mean that the global existence of the solution u is equivalent to the existence of an *a priori* estimate of $\|u(t)\|$ on $[0, T(x))$. In applications, we establish such *a priori* estimates by standard methods in the theory of partial differential equations (multipliers, comparison principles, and maximum principles), as well as by various techniques involving differential or integral inequalities more specific to evolution equations (first integrals, Liapunov functions, and variants of Gronwall's lemma).

Proof of Theorem 4.3.4. It is clear that (4.10) implies that if $T(x) < \infty$, then $\|u(t)\| \to \infty$ as $t \uparrow T(x)$. Let $x \in X$. We set

$$T(x) = \sup\{T > 0; \exists u \in C([0, T], X) \text{ solution of (4.9)}\}.$$

By Proposition 4.3.3, we know that $T(x) > 0$. On the other hand, the uniqueness (Lemma 4.3.2) allows us to build a maximal solution $u \in C([0, T(x)), X)$ of (4.9). It remains to show (4.10). Inequality (4.10) being immediate if $T(x) = \infty$, we may assume that $T(x) < \infty$. We argue by contradiction, assuming that there exists $t \in [0, T(x))$ such that (4.10) does not hold. We then have

$$T(x) - t < T_M,$$

with $M = \|u(t)\|$. Let $v \in C([0, T_M], X)$ be the solution, given by Proposition 4.3.3, of

$$v(s) = \mathcal{T}(s)u(t) + \int_0^s \mathcal{T}(s - \sigma)F(v(\sigma))\,d\sigma,$$

for all $s \in [0, T_M]$. We then define $w \in C([0, t + T_M], X)$ by

$$w(s) = \begin{cases} u(s), & \text{if } s \in [0, t]; \\ v(s - t), & \text{if } s \in [t, t + T_M]. \end{cases}$$

We verify easily that w is a solution of (4.9) with $T = t + T_M$, which contradicts the definition of $T(x)$, since $t + T_M > T(x)$. □

Remark 4.3.6. It may very well happen that, for the same equation, $T(x) < \infty$ for some initial data, and $T(x) = \infty$ for others. For example, choose $X = \mathbb{R}$, $A = 0$, and $F(u) = u^3 - u$. This choice corresponds to the ordinary differential equation $u' = u^3 - u$. If $|x| \leq 1$ we have $T(x) = \infty$, and if $|x| > 1$ we have $T(x) < \infty$. In the last case, (4.10) gives $12|u(t)|^2 \geq (T(x) - t)^{-1} - 4$. This estimate describes the blow-up phenomenon sharply, since the solutions actually blow up as $(T(x) - t)^{-\frac{1}{2}}$.

4.3.2. Continuous dependence on initial data

Proposition 4.3.7. *Following the notation of Theorem 4.3.4, we have the following properties:*

(i) $T : X \to (0, \infty]$ *is lower semicontinuous;*

(ii) *if $x_n \to x$ and if $T < T(x)$, then $u_n \to u$ in $C([0,T], X)$, where u_n and u are the solutions of (4.9) corresponding to the initial data x_n and x.*

Proof. Let $x \in X$, and let u be the solution of (4.9) given by Theorem 4.3.4. Let $0 < T < T(x)$. It suffices to show that if $(x_n)_{n \geq 0} \subset X$ is a sequence such that $x_n \longrightarrow x$ as $n \to \infty$, then for n sufficiently large $T(x_n) > T$ and $u_n \to u$ in $C([0,T], X)$. To see this, set

$$M = 2 \sup_{t \in [0,T]} \|u(t)\|,$$

and

$$\tau_n = \sup\{t \in [0, T(x_n)); \|u_n(s)\| \leq 2M, \forall s \in [0,t]\}.$$

For n large enough, we have $\|x_n\| < M$; and so $\tau_n > T_M > 0$. For all $t \leq T$, $t \leq \tau_n$, we have

$$\|u(t) - u_n(t)\| \leq \|x - x_n\| + L(2M) \int_0^t \|u(s) - u_n(s)\| \, ds;$$

and it follows from Lemma 4.2.1 that

$$\|u(t) - u_n(t)\| \leq \|x - x_n\| e^{TL(2M)}, \tag{4.11}$$

for $t \leq T$, $t \leq \tau_n$. In particular, we deduce from (4.11) that for n large enough we have

$$\|u_n(t)\| \leq M,$$

for $t \leq \min\{T, \tau_n\}$; and so $\tau_n > \min\{T, \tau_n\}$, i.e. $\tau_n > T$. We then have $T(x_n) > T$. Applying (4.11) again, we see that $u_n \to u$ in $C([0,T], X)$. This completes the proof. $\square$

Remark 4.3.8. Actually, T may be discontinuous. For example, choose $X = \mathbb{R}^2$, $A = 0$, and $F(u,v) = (vu^2, -2)$. For $x = (1, 2)$ we have $T(x) = 1$ and the corresponding solution is $((1-t)^{-2}, 2(1-t))$. For $x_\varepsilon = ((1+\varepsilon)^{-1}, 2)$ we have $T(x_\varepsilon) = \infty$ and the corresponding solution is $((\varepsilon + (1-t)^2)^{-1}, 2(1-t))$.

4.3.3. Regularity

In some cases, it is possible to give a more precise result on the regularity of solutions of (4.9). In particular, we have the following.

Proposition 4.3.9. *Assume that X is reflexive. Let $T > 0$, $x \in X$, and let $u \in C([0,T], X)$ be a solution of problem (4.9). Then, if $x \in D(A)$, u is the solution of problem (4.6)–(4.8).*

Proof. Let $h > 0$ and let $t \in [0, T - h]$. It is easy to see that

$$u(t+h) - u(t) = T(h)x - x + \int_0^t T(s)\{F(u(t+h-s)) - F(u(t-s))\}\,\mathrm{d}s$$
$$+ \int_0^h T(t+s)F(u(s))\,\mathrm{d}s.$$

Hence,

$$\|u(t+h) - u(t)\| \leq \|T(h)x - x\| + h \sup_{s \in [0,T]} \|F(u(s))\|$$
$$+ L(M) \int_0^t \|u(s+h) - u(s)\|\,\mathrm{d}s.$$

Furthermore, we have

$$T(h)x - x = \int_0^h T(s)Ax\,\mathrm{d}s;$$

and so $\|T(h)x - x\| \leq h\|Ax\|$. Applying Lemma 4.2.1, we obtain

$$\|u(t+h) - u(t)\| \leq Ch,$$

for $0 \leq t < t + h \leq T$. Consequently, $u : [0, T] \to X$ is Lipschitz continuous, and it follows that $F(u) : [0, T] \to X$ is also Lipschitz continuous. We conclude by applying Corollary 1.4.41 and Proposition 4.1.6. □

Remark 4.3.10. If X is not reflexive, the conclusion of Proposition 4.3.9 may fail, as shown by the following example. Choose $X = C_0(\mathbb{R}) \times C_0(\mathbb{R})$, where $C_0(\mathbb{R})$ is the space of functions of $C(\mathbb{R})$ which dies away to 0 as $x \to \pm\infty$, equipped with the norm $L^\infty(\mathbb{R})$. We define the operator A by

$$\begin{cases} D(A) = \{(u, v) \in X \cap C^1(\mathbb{R}^2);\ (u', v') \in X\}; \\ A(u, v) = (u', v'), \quad \forall (u, v) \in D(A). \end{cases}$$

A is m-dissipative with dense domain, and generates the semigroup $(T(t))_{t\geq 0}$ given by

$$T(t)(u,v) = (u(t+\cdot), v(t+\cdot)),$$

for $t \geq 0$, $x \in \mathbb{R}$. Next, consider the Lipschitz continuous function $F : X \to X$ given by

$$F(u,v) = (v^+, 0), \quad \forall (u,v) \in X.$$

For all $(x,y) \in X$, the corresponding solution (u,v) of (4.9) is given by

$$(u,v)(t) = (x(t+\cdot) + ty^+(t+\cdot), y(t+\cdot)).$$

Taking $(x,y) \in D(A)$ such that $y(0) = 0$ and $y'(0) \neq 0$, y^+ is not in $C^1(\mathbb{R})$, and so $(u,v)(t) \notin D(A)$, for $t \neq 0$.

4.4. Isometry groups

In the case in which A generates an isometry group (see Theorem 3.2.3), and in particular when X is a Hilbert space and A is skew-adjoint, we can also solve (4.7) for $t \leq 0$. Indeed, solving the problem

$$\begin{cases} u \in C([-T,0], X) \cap C^1([-T,0], \overline{X}); \\ u'(t) = \overline{A}u(t) + F(u(t)), \quad \forall t \in [-T, 0]; \\ u(0) = x; \end{cases}$$

is equivalent to solving

$$\begin{cases} v \in C([0,T], X) \cap C^1([0,T], \overline{X}); \\ v'(t) = -\overline{A}u(t) - F(u(t)), \quad \forall t \in [0, T]; \\ v(0) = x; \end{cases}$$

setting $u(t) = v(-t)$, for $t \in [-T, 0]$. The second problem is solved by Theorem 4.3.4, since $-A$ is m-dissipative and $-F$ is Lipschitz continuous on bounded sets of X.

Notes. One finds generalizations of the results of §4.3 in Segal [1] and Weissler [1]. Also consult Ball [1, 2] for an interesting discussion about the blow-up phenomenon.

5
The heat equation

Throughout this chapter, we assume that Ω is a bounded subset of $\mathbb{R}^N$ with Lipschitz continuous boundary, and we use the notation of §3.5.1. In particular, $X = C_0(\Omega)$ and $Y = L^2(\Omega)$. In addition, we consider a locally Lipschitz continuous function $g \in C(\mathbb{R}, \mathbb{R})$, such that

$$g(0) = 0.$$

We define the function $F : X \to X$ by

$$F(u)(x) = g(u(x)),$$

for all $u \in X$ and $x \in \Omega$. It is easy to check that F is Lipschitz continuous on bounded sets of X. In what follows, we denote g and F by the same expression.

5.1. Preliminaries

Given $\varphi \in X$, we look for $T > 0$ and u solving the problem

$$\begin{cases} u \in C([0,T], X) \cap C((0,T], H_0^1(\Omega)) \cap C^1((0,T], L^2(\Omega)); \\ \Delta u \in C((0,T], L^2(\Omega)); & (5.1) \\ u_t - \Delta u = F(u), \quad \forall t \in (0,T]; & (5.2) \\ u(0) = \varphi. & (5.3) \end{cases}$$

The result is the following.

Proposition 5.1.1. *Let $\varphi \in X$, $T > 0$, and let $u \in C([0,T], X)$. Then u is solution of (5.1)–(5.3) if and only if u satisfies*

$$u(t) = T(t)\varphi + \int_0^t T(t-s)F(u(s))\,ds, \qquad (5.4)$$

for all $t \in [0,T]$.

Proof. Let u be a solution of (5.1)–(5.3), let $t \in (0,T]$, and let $\varepsilon \in (0,t]$. We set

$$v(s) = u(\varepsilon + s),$$

for $0 \leq s \leq t - \varepsilon$. It is clear that v is a solution of (5.2) on $[0, t - \varepsilon]$ and that $v(0) = u(\varepsilon) \in D(B)$. Hence, we have (Lemma 4.1.1)

$$v(s) = S(s)u(\varepsilon) + \int_0^s S(s-\sigma)F(v(\sigma))\,d\sigma,$$

for all $s \in [0, t - \varepsilon]$. Applying Lemma 3.5.6, we deduce that

$$u(s+\varepsilon) = T(s)u(\varepsilon) + \int_0^s T(s-\sigma)F(u(\sigma+\varepsilon))\,d\sigma,$$

for all $s \in [0, t - \varepsilon]$. Since $u \in C([0,T], X)$, we have, for all $s \in [0, t)$,

$$T(s)u(\varepsilon) \to T(s)\varphi,$$

as $\varepsilon \downarrow 0$;

$$F(u(\cdot + \varepsilon)) \to F(u(\cdot)),$$

uniformly on $[0, s]$ as $\varepsilon \downarrow 0$. Letting first $\varepsilon \downarrow 0$, and then $s \uparrow t$, we deduce (5.4).

Conversely, let $u \in C([0,T], X)$ be a solution of (5.4). We consider $0 < t \leq T$. By Proposition 3.5.2, we have $T(t)\varphi \in H_0^1(\Omega)$, and the function $s \mapsto T(t-s)F(u(s))$ belongs to $C([0,t), H_0^1(\Omega))$, with

$$\|T(t-s)F(u(s))\|_{H^1} \leq C(1 + (t-s)^{-1/2}) \in L^1(0,t);$$

and so (Proposition 1.4.14 and Corollary 1.4.23) $u(t) \in H_0^1(\Omega)$. A similar estimate shows that actually $u \in C((0,T], H_0^1(\Omega))$,

$$\|u(t)\|_{H^1} \leq C(1 + t^{-1/2}).$$

Since g is Lipschitz continuous on bounded subsets of $\mathbb{R}$ and since the range of u is bounded, we conclude (Proposition 1.3.5) that $F(u(t)) \in H_0^1(\Omega)$, and that

$$\|F(u(t))\|_{H^1} \leq C(1 + t^{-1/2}).$$

It follows that $F(u)$ is weakly continuous as a map from $(0, T]$ to $H_0^1(\Omega)$. Take $0 < t < T$ again. Applying Proposition 3.5.2 again, we obtain $\Delta T(t)\varphi \in L^2(\Omega)$, and that the function $s \mapsto \Delta T(t-s)F(u(s))$ is weakly continuous as a map from $(0,t)$ to $L^2(\Omega)$, with

$$\|DT(t-s)F(u(s))\|_{L^2} \leq C(t-s)^{-1/2}(1 + s^{-1/2}) \in L^1(0,t);$$

and so $\Delta u(t) \in L^2(\Omega)$. Consequently, $u(t) \in D(B)$, for all $t \in (0,T]$. We show similarly that $u \in C((0,T), D(B))$. We then set

$$v(s) = u(\varepsilon + s),$$

for $0 \leq s \leq t - \varepsilon$. We have

$$v(s) = S(s)u(\varepsilon) + \int_0^s S(s-\sigma)F(v(\sigma))\,ds,$$

for all $s \in [0, t - \varepsilon]$. We conclude by applying Corollary 4.1.8. □

Note that here it is not possible to invoke Proposition 4.3.9, since X is not reflexive.

Remark 5.1.2. Considering the above estimates in more detail, we obtain, as a consequence of Proposition 3.5.2,

$$\|\nabla u(t)\|_{L^2} \leq C|\Omega|^{1/2}(t^{-1/2} + t^{1/2});$$
$$\|\Delta u(t)\|_{L^2} \leq C|\Omega|^{1/2}(t^{-1} + t);$$

where C depends only on g and $\sup_{0 \leq t \leq T} \|u(t)\|$.

Remark 5.1.3. Similarly, we easily verify that if $\varphi \in H_0^1(\Omega)$, then we have

$$\|\nabla u(t)\|_{L^2} \leq C|\Omega|^{1/2}(1 + t^{1/2});$$
$$\|\Delta u(t)\|_{L^2} \leq C|\Omega|^{1/2}(t^{-1/2} + t);$$

where C depends only on g, $\sup_{0 \leq t \leq T} \|u(t)\|$ and $\|\varphi\|_{H^1}$.

Remark 5.1.4. The same method also shows that if we assume further that $\varphi \in D(B)$, then

$$\|\Delta u(t)\|_{L^2} \leq C|\Omega|^{1/2}(1 + t),$$

where C depends only on g, $\sup_{0 \leq t \leq T} \|u(t)\|$, and $\|\varphi\|_{H^1}$.

5.2. Local existence

Applying Proposition 5.1.1 and Theorem 4.3.4, we obtain the following result.

Theorem 5.2.1. For all $\varphi \in X$, there exists a unique function u, defined on a maximal interval $[0, T(\varphi))$, which is a solution of (5.1)–(5.3) for all $T \in (0, T(\varphi))$. In addition, if $T(\varphi) < \infty$, then $\|u(t)\| \to \infty$ as $t \uparrow T(\varphi)$.

Remark 5.2.2. u depends continuously on φ (this follows from Proposition 4.3.7).

The following proposition gives improved regularity of u if φ is more regular.

Proposition 5.2.3. *Assume that $\varphi \in X \cap H_0^1(\Omega)$. Then the solution u corresponding to (5.1)–(5.3) is in $C([0,T(\varphi)), H_0^1(\Omega))$. Suppose further that $\triangle \varphi \in L^2(\Omega)$; then $u \in C([0,T(\varphi)), D(B)) \cap C^1([0,T(\varphi)), L^2(\Omega))$.*

Proof. Assume that $\varphi \in X \cap H_0^1(\Omega)$, and let $t \in (0, T(\varphi))$. Applying (5.2), Proposition 3.16, and (3.31), we obtain

$$\|u(t) - \varphi\|_{H^1} \leq \|S(t)\varphi - \varphi\|_{H^1} + C \int_0^t \frac{1}{\sqrt{t-s}} \|F(u(s))\| ds$$

$$\leq \|S(t)\varphi - \varphi\|_{H^1} + C\sqrt{t} \longrightarrow 0 \quad \text{as } t \downarrow 0.$$

Therefore $u \in C([0, T(\varphi)), H_0^1(\Omega))$. In particular, if $T < T(\varphi)$, then u is bounded in $H_0^1(\Omega)$ on $[0,T]$, and then so is $F(u)$. Consequently, in the case in which $\triangle \varphi \in L^2(\Omega)$, it follows from (5.2) and (3.32) that

$$\|\triangle(u(t) - \varphi)\|_{L^2} \leq \|(S(t) - I)\triangle\varphi\|_{L^2} + C \int_0^t \frac{1}{\sqrt{t-s}} \|F(u(s))\|_{H^1} ds \xrightarrow[t\downarrow 0]{} 0;$$

and so $u \in C([0, T(\varphi)), D(B))$. In particular, $u_t(t) \longrightarrow \triangle\varphi + F(\varphi)$ in $L^2(\Omega)$ as $t \downarrow 0$. Furthermore, for $t < T(\varphi)$, we have

$$\frac{u(t) - \varphi}{t} = \frac{S(t) - I}{t}\varphi + \frac{1}{t}\int_0^t S(t-s)F(u(s))\, ds \longrightarrow \triangle\varphi + F(\varphi) \quad \text{as } t \downarrow 0.$$

Consequently, $(d^+u/dt)(0) = \lim_{t\downarrow 0} u_t(t)$; and so $u \in C^1([0, T(\varphi)), L^2(\Omega))$. □

5.3. Global existence

We establish here two kinds of results. First, we show that if g satisfies certain conditions for $|x|$ large, then all solutions of (5.1)–(5.3) are global. Then, in another spirit, some results prove that if g satisfies certain conditions for $|x|$ small, then the solutions of (5.1)–(5.3) with small initial data are global. We begin with the following result (the maximum principle).

Proposition 5.3.1. *Let $T > 0$ and $\varphi \in X$. Let $u \in C([0,T], X) \cap C((0,T), H_0^1(\Omega)) \cap C^1((0,T), L^2(\Omega))$ with $\triangle u \in C((0,T), L^2(\Omega))$, and $f \in C([0,T], X)$ be such that*

$$\begin{cases} u_t - \triangle u = f, & \forall t \in (0,T); \\ u(0) = \varphi. \end{cases} \quad (5.5)$$

Assume further that there exists a constant C such that

$$|f(t,x)| \leq C|u(t,x)|, \quad (5.6)$$

in $[0,T] \times \Omega$. Then, if $\varphi \geq 0$, we have $u(t) \geq 0$ for all $t \in [0,T]$.

Proof. For $t \in (0, T)$, we set
$$v(t) = \int (u^-(t))^2.$$
Multiply (5.5) by $-u^-(t)$ and integrate over Ω. Integrating by parts and applying (5.6), we obtain (see the proof of Lemma 3.5.9)
$$v'(t) \leq \int |f(t)| u^-(t) \leq C \int |u(t)| u^-(t) = Cv(t).$$
Integrating the last inequality, we obtain, for all $0 < s \leq t < T$,
$$v(t) \leq v(s) e^{C(t-s)}.$$
Letting $s \downarrow 0$, it follows that
$$v(t) \leq e^{Ct} \int (\varphi^-)^2 = 0, \quad \forall t \in (0, T);$$
hence the result. $\square$

Remark 5.3.2. Applying Proposition 5.3.1 with $v = -u$, we see that if $\varphi \leq 0$, then we have $u(t) \leq 0$ for all $t \in [0, T]$.

We give a first result of global existence.

Proposition 5.3.3. *Assume that there exist $K, C < \infty$ such that*
$$xg(x) \leq C|x|^2, \tag{5.7}$$
for $|x| \leq K$. Then, for all $\varphi \in X$, the solution of (5.1)–(5.3) given by Theorem 5.2.1 is global.

Proof. We proceed in two steps.

Step 1. Assume first that $\varphi \geq 0$. It is easy to verify that, for all $T < T(\varphi)$, $h(t) = F(u(t))$ satisfies (5.6). Then we have $u(t) \geq 0$ for all $t \in [0, T(\varphi))$. Furthermore, by (5.7) we have
$$g(x) \leq Cx,$$
for $x \geq K$. Since g is bounded on $[0, K]$ and by possibly modifying C, it follows that
$$g(x) \leq C + Cx,$$
for $x \geq 0$; and so
$$\|F(u(t))^+\| \leq C + C\|u(t)\|,$$
for all $t \in [0, T(\varphi))$.

Applying (5.4), Lemma 3.5.9, and Gronwall's lemma, we deduce that

$$\|u(t)\| \leq (C + \|\varphi\|)e^{Ct}, \tag{5.8}$$

for all $t \in [0, T(\varphi))$; and so (Theorem 5.2.1) $T(\varphi) = \infty$.

Step 2. In the general case, we set $\psi = |\varphi|$ and we denote by v the corresponding solution of (5.1)–(5.3). Let $T < \min\{T(\psi), T(\varphi)\} = T(\varphi)$. We easily verify that $w = v - u$ fulfills the assumptions of Proposition 5.3.1; hence

$$u(t) \leq v(t),$$

for all $t \in [0, T]$. We then use $z = \underline{v} + u$, where $\underline{v}$ is the maximal solution of the equation

$$\underline{v} = \mathcal{T}(t)\psi + \int_0^t \mathcal{T}(t-s)\{-g(-\underline{v})\}\,ds.$$

Since $-g(-x)$ verifies the same assumptions as $g(x)$, $\underline{v}$ is a global solution (Step 1). Applying Proposition 5.3.1, we obtain that

$$u(t) \geq -\underline{v}(t),$$

for all $t \in [0, T]$; and so

$$\|u(t)\| \leq \max\{\|v(t)\|, \|\underline{v}(t)\|\},$$

for all $t \in [0, T]$. It follows that $T(\varphi) \geq T(\psi) = \infty$. This completes the proof. $\square$

Remark 5.3.4. If $C = 0$ in (5.7), it follows from (5.8) that all solutions of (5.1)–(5.3) satisfy

$$\|u(t)\| \leq K + \|\varphi\|,$$

for $t \geq 0$. Actually, this result can be sharpened by the following proposition, whose proof is quite simple.

Proposition 5.3.5. *If $C = 0$ in (5.7), then solutions of (5.1)–(5.3) satisfy*

$$\|u(t)\| \leq \max\{K, \|\varphi\|\},$$

for $t \geq 0$.

68 *The heat equation*

Proof. As in Proposition 5.3.3, we may restrict ourselves to the case $\varphi \geq 0$. Set $k = \max\{K, \|\varphi\|\}$ and multiply (5.2) by $(u-k)^+ \in H_0^1(\Omega)$. Integrating by parts and setting $f(t) = \int ((u(t)-k)^+)^2$, it follows that

$$f'(t) \leq C \int_\Omega g(u(t))(u(t)-k)^+ \leq 0.$$

We conclude that $f(t) \leq f(0) = 0$, and so $u \leq k$; hence the result. □

If the constant C of (5.7) is sufficiently small, we still have a result of this kind. To see this, we consider λ given by (2.2).

Proposition 5.3.6. *If $C < \lambda$ in (5.7), then, for all $\varphi \in X$, the solution u of (5.1)–(5.3) satisfies*

$$\sup_{0 \leq t < \infty} \|u(t)\| < \infty.$$

Proof. We have $xg(x) \leq C_1 + Cx^2$. As in Proposition 5.3.3, we may assume that $\varphi \geq 0$. Let $\varphi \geq 0$ and let u be the corresponding solution. We multiply (5.2) by u. Integrating by parts and setting $f(t) = \int u(t)^2$, it follows that

$$f'(t) \leq -2\lambda f(t) + 2\int u(t)g(u(t)) \leq -2(\lambda - C)f(t) + C_1|\Omega|.$$

By Lemma 8.4.6 (see below), we have

$$f(t) \leq f(0) + C_1 \frac{|\Omega|}{\lambda - C};$$

and thus $\sup_{0 \leq t < \infty} \|u(t)\|_{L^2} = K < \infty$. Let $p \in (N/2, \infty)$. By (3.34) and (3.37), we have

$$\|T(t)\|_{\mathcal{L}(L^p, L^\infty)} \in L^1(0, +\infty).$$

We then note that there exists a constant C_2 (see the proof of Proposition 5.3.3) such that

$$\|g(u(t))^+\|_{L^p} \leq C_2 + C\|u(t)\|_{L^p},$$

for all $t \geq 0$; and so

$$\|u(t)\| \leq \|\varphi\| + \|T(\cdot)\|_{\mathcal{L}(L^p, L^\infty)}(C_2 + C \sup_{0 \leq s \leq t} \|u(s)\|_{L^p})$$
$$\leq C_3 + C_4 \sup_{0 \leq s \leq t} \|u(s)\|_{L^p},$$

for all $t \geq 0$. Finally, invoking Hölder's inequality, for all $\varepsilon > 0$, there exists $C(\varepsilon)$ such that

$$\|v\|_{L^p} \leq \varepsilon \|v\| + C(\varepsilon)\|v\|_{L^2},$$

for all $v \in X$. Choosing ε such that $\varepsilon C_4 \leq 1/2$, it follows that

$$\|u(t)\| \leq C_3 + KC(\varepsilon)C_4 + \frac{1}{2}\sup_{0 \leq s \leq t}\|u(s)\|,$$

for all $t \geq 0$. Therefore

$$\sup_{0 \leq s \leq t}\|u(s)\| \leq 2C_3 + 2KC(\varepsilon)C_4,$$

for all $t \geq 0$; hence the result. □

Now we show that if g fulfills certain conditions in a neighbourhood of 0, then solutions with small initial data are global. To see this, consider λ given by (2.2).

Proposition 5.3.7. *Suppose that there exists $\alpha > 0$ and $\mu < \lambda$ such that*

$$xg(x) \leq \mu|x|^2, \quad \text{for } |x| \leq \alpha.$$

Then, there exists $A < \infty$ such that, if $\|\varphi\| < \alpha A$, the corresponding solution u of (5.1)–(5.3) is global and satisfies

$$\|u(t)\| \leq A\|\varphi\|e^{-(\lambda-\mu)t},$$

for $t \geq 0$.

Proof. As in Proposition 5.3.3, it suffices to deal with the case $\varphi \geq 0$. Set

$$T = \sup\{t \in [0, T(\varphi)); \|u(s)\| \leq \alpha \text{ for } s \in [0,t]\} > 0.$$

Multiply (5.2) by u. Integrating by parts, and letting $f(t) = \int u(t)^2$, for $t \in [0,T]$, it follows that

$$f'(t) \leq -2\lambda f(t) + 2\int u(t)g(u(t)) \leq -2(\lambda - C)f(t);$$

and so

$$\|u(t)\|_{L^2} \leq \|\varphi\|_{L^2}e^{-(\lambda-\mu)t},$$

for all $t \in [0,T]$. We write

$$e^{(\lambda-\mu)t}u(t) = e^{(\lambda-\mu)t}T(t)\varphi + \int_0^t e^{(\lambda-\mu)(t-s)}T(t-s)\left(e^{(\lambda-\mu)s}F(u(s))\right)ds.$$

Note that (see the proof of Proposition 5.3.6)

$$\|e^{(\lambda-\mu)t}T(t)\|_{\mathcal{L}(L^p,L^\infty)} \in L^1(0,+\infty).$$

Consequently, arguing as in the proof of Proposition 5.3.6, we obtain

$$e^{(\lambda-\mu)t}\|u(t)\| \leq e^{-\mu t}\|\varphi\| + C\|\varphi\|_{L^2} + \frac{1}{2}\sup_{0\leq s\leq t} e^{(\lambda-\mu)s}\|u(s)\|,$$

for all $t \in [0,T]$. Hence, there exists A such that

$$\sup_{0\leq s\leq t} e^{(\lambda-\mu)s}\|u(s)\| \leq A\|\varphi\|,$$

for all $t \in [0,T]$. It follows that if $\|\varphi\| < \alpha A$, then $T = \infty$, which completes the proof. □

Remark 5.3.8. We see that if $\mu \downarrow 0$, we may take $\delta \uparrow \lambda$. If g has a higher order near 0, we have a more precise result.

Proposition 5.3.9. *Assume that there exist $\mu \geq 0$, $\varepsilon > 0$, and $\alpha > 0$ such that*

$$xg(x) \leq \mu|x|^{2+\varepsilon}, \quad \text{for } |x| \leq \alpha.$$

Then there exist $\beta, \gamma > 0$ such that, if $\|\varphi\| < \beta$, then the corresponding solution u of (5.1)–(5.3) is global and

$$\|u(t)\| \leq \gamma\|\varphi\|e^{-\lambda t},$$

for $t \geq 0$.

Proof. As in Proposition 5.3.3, we may assume that $\gamma \geq 0$. Set

$$\theta(x) = \frac{\mu M}{\varepsilon\lambda}x^{1+\varepsilon} - x,$$

for $x \geq 0$, and $-\xi = \min\theta < 0$. For all $a \in (0,\xi)$, there exist $0 < x_a < y_a$ such that

$$\theta(x_a) + a = \theta(y_a) + a = 0.$$

Furthermore, we have $a < x_a < a(1+\varepsilon)/\varepsilon$ (see Figure 5.1).

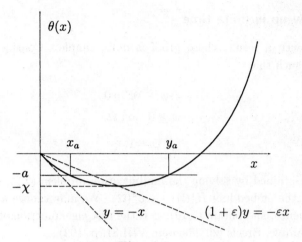

Fig. 5.1

Suppose that $\|\varphi\| < \alpha$ and set

$$T = \sup\{t \in [0, T(\varphi)), \|u(s)\| \leq \alpha \text{ for } s \in [0,t]\} > 0,$$
$$f(t) = \sup_{0 \leq s \leq t} \|u(s)\|.$$

We have

$$\|F(u(t))^+\| \leq \mu \|u(t)\|^{1+\varepsilon},$$

for $t \in [0, T]$. Applying (5.4), (3.37), and Lemma 3.5.9, it follows that

$$e^{\lambda t}\|u(t)\| \leq M\|\varphi\| + \mu M \int_0^t e^{-\varepsilon \lambda s}\{e^{\lambda s}\|u(s)\|\}^{1+\varepsilon}\,ds \leq M\|\varphi\| + \frac{\mu M}{\varepsilon \lambda} f(t)^{1+\varepsilon}.$$

Thus,

$$\theta(f(t)) + M\|\varphi\| \geq 0,$$

for all $t \in [0, T)$. If we assume further that $M\|\varphi\| < \xi$, then this implies that $f(t) \in [0, x_{M\|\varphi\|}) \cup (y_{M\|\varphi\|}, \infty)$. Since $f(0) \in [0, x_{M\|\varphi\|})$ and $f(t)$ is continuous in t, we have

$$f(t) \in [0, x_{M\|\varphi\|}),$$

for all $t \in [0, T)$. We conclude as in Proposition 5.3.7, and we obtain the result with $\beta = \min\{\alpha, \xi/M\}$ and $\gamma = (1+\varepsilon)M/\varepsilon$. □

5.4. Blow-up in finite time

We begin with a result whose proof is quite simple. Consider the function $\psi \in D(B)$, such that

$$\Delta \psi + \lambda \psi = 0, \qquad (5.9)$$
$$\psi \geq 0 \quad \text{on } \Omega, \qquad (5.10)$$
$$\int_\Omega \psi = 1. \qquad (5.11)$$

ψ is easily obtained by solving the minimization problem (2.2), using the compactness of the embedding $H_0^1(\Omega) \hookrightarrow L^2(\Omega)$. We may choose a positive minimizing sequence, and obtain (5.10). ψ is the first eigenfunction of $-\Delta$ in $H_0^1(\Omega)$ (see, for example, Brezis [2], Theorem VIII.31, p. 192).

Proposition 5.4.1. *Suppose that there exist $\alpha, \beta, \varepsilon > 0$ such that $g(x) \geq \alpha x^{1+\varepsilon} - \beta x$, for $x \geq 0$. Let $\varphi \in X$, $\varphi \geq 0$ on Ω, be such that*

$$\int \varphi \psi \geq \left(\frac{\lambda + \beta}{\alpha}\right)^{\frac{1}{1+\varepsilon}}.$$

Then $T(\varphi) < \infty$.

Proof. Denote by u the solution of (5.1)–(5.3) given by Theorem 5.2.1. By Proposition 5.3.1, we have $u(t) \geq 0$ on Ω, for all $t \in [0, T(\varphi))$. Set

$$f(t) = \int_\Omega u(t)\psi \geq 0,$$

for $t \in [0, T(\varphi))$. Applying (5.9) and Lemma 2.34, we obtain, for all $t \in [0, T(\varphi))$,

$$f'(t) = \int u_t(t)\psi = \int (\Delta u(t) + g(u(t)))\psi$$
$$= \int u(t)\Delta\psi + \int g(u(t))\psi = -\lambda f(t) + \int g(u(t))\psi$$
$$\geq -(\lambda + \beta)f(t) + \alpha \int u(t)^{1+\varepsilon}\psi.$$

On the other hand, by (5.11) and Hölder's inequality,

$$f(t) \leq \left(\int_\Omega u(t)^{1+\varepsilon}\psi\right)^{\frac{1}{1+\varepsilon}} \left(\int_\Omega \psi\right)^{\frac{\varepsilon}{1+\varepsilon}} \leq \left(\int_\Omega u(t)^{1+\varepsilon}\psi\right)^{\frac{1}{1+\varepsilon}};$$

and so

$$f'(t) \geq f(t)(-(\lambda+\beta) + \alpha f(t)^\varepsilon), \qquad (5.12)$$

for all $t \in (0, T(\varphi))$. Let $T = \sup\{t \in (0, T(\varphi)), f' > 0 \text{ on } (0,t)\} > 0$. If $T < T(\varphi)$, we have $f'(T) = 0$ and $f(T) > f(0)$, which contradicts (5.12). Thus, we have $T = T(\varphi)$ and $f' > 0$ on $[0, T(\varphi))$. Now let $\delta > 0$ be such that

$$(\alpha - \delta) f(0)^\varepsilon = \lambda + \beta.$$

We deduce from (5.12) that, for all $t \in (0, T(\varphi))$,

$$\begin{aligned} f'(t) &\geq \delta f(t)^{1+\varepsilon} + f(t)(-(\lambda + \beta) + (\alpha - \delta) f(t)^\varepsilon) \\ &\geq \delta f(t)^\varepsilon + f(t)(-(\lambda + \beta) + (\alpha - \delta) f(0)^\varepsilon) \geq \delta f(t)^{1+\varepsilon}, \end{aligned}$$

i.e.

$$-\left(\frac{f(t)^{-\varepsilon}}{\varepsilon}\right)' \geq (\delta t)'.$$

From this, we easily deduce that

$$0 \leq \frac{f(t)^{-\varepsilon}}{\varepsilon} \leq \frac{f(0)^{-\varepsilon}}{\varepsilon} - \delta t,$$

for all $t \in (0, T(\varphi))$; and so $\varepsilon \delta T(\varphi) \leq f(0)^{-\varepsilon}$; hence the result. □

Remark 5.4.2. It is important to note that the above argument only shows that $\varepsilon \delta T(\varphi) \leq f(0)^{-\varepsilon}$ and not that $\varepsilon \delta T(\varphi) = f(0)^{-\varepsilon}$. For further discussion concerning this question, see Ball [1, 2].

Remark 5.4.3. If we take $\zeta \in X$ such that $\zeta \geq 0$, then for $k > 0$ large enough $\varphi = k\zeta$ satisfying the assumptions of Proposition 5.4.1, and so $T(\varphi) < \infty$.

Now we show a second blow-up result, using a different method. We need the functional E defined by

$$E(u) = \frac{1}{2} \int |\nabla u|^2 - \int G(u),$$

for $u \in X \cap H_0^1(\Omega)$, where

$$G(x) = \int_0^x g(s) \, ds,$$

for $x \in \mathbb{R}$.

Proposition 5.4.4. Assume that there exists $K \geq 0$ and $\varepsilon > 0$ such that

$$xg(x) \geq (2 + \varepsilon) G(x),$$

for $|x| \geq K$. Set $\mu = \min_{0 \leq x \leq K} xg(x)$ and $\nu = \max_{0 \leq x \leq K} G(x)$. Let $\varphi \in X \cap H_0^1(\Omega)$ be such that

$$(2+\varepsilon)E(\varphi) < |\Omega|(\mu - 2\varepsilon\nu).$$

Then $T(\varphi) < \infty$.

The proof makes use of the following two lemmas.

Lemma 5.4.5. Let $T > 0$ and $u \in C((0,T), X) \cap C((0,T), D(B)) \cap C^1((0,T), L^2(\Omega))$. Then

$$\int_s^t \int_\Omega (\Delta u + g(u))u_t + E(u(t)) = E(\varphi),$$

for all $0 < s < t < T$.

Proof. By density, we may restrict ourselves to the case in which $u \in C^1((0,T), D(A))$ (by possibly replacing u by $\frac{1}{h}\int_t^{t+h} J_\lambda u(s)\,ds$). Then we have

$$\int_\Omega (\Delta u + g(u))u_t = \int_\Omega (-\nabla u \cdot \nabla u_t + g(u)u_t) = \frac{d}{dt}E(u(t)),$$

and hence the result. □

Lemma 5.4.6. Let $\varphi \in X$ and let u be the corresponding solution to (5.1)–(5.3). Then

$$\frac{d}{dt}\int_\Omega u(t)^2\,dx + 2\int_\Omega |\nabla u(t)|^2\,dx = 2\int_\Omega u(t)g(u(t))\,dx; \tag{5.13}$$

$$\int_s^t \int_\Omega u_t^2 + E(u(t)) = E(u(s)); \tag{5.14}$$

for all $0 < s < t < T(\varphi)$.

Proof. (5.13) is obtained by multiplying (5.2) by u_t, integrating over Ω, and then integrating by parts.

We obtain (5.14) by applying Lemma 5.4.5 and (5.2). □

Proof of Proposition 5.4.4. Set $f(t) = \int_\Omega u(t)^2$. By (5.13), for $t \in (0, T(\varphi))$, we have

$$f'(t) = -2\int_\Omega |\nabla u(t)|^2 + 2\int_{\{|u|<K\}} u(t)g(u(t)) + 2\int_{\{|u|>K\}} u(t)g(u(t))$$

$$\geq -2\int_\Omega |\nabla u(t)|^2 + 2\int_{\{|u|<K\}} u(t)g(u(t)) + 2(2+\varepsilon)\int_{\{|u|>K\}} G(u(t)). \tag{5.15}$$

On the other hand, observe that, by Proposition 5.2.3, we may let $s = 0$ in (5.14); and so

$$2(2+\varepsilon) \int_{\{|u| \geq K\}} G(u(t)) = -2(2+\varepsilon) \int_{\{|u| < K\}} G(u(t))$$
$$+(2+\varepsilon) \int_\Omega |\nabla u(t)|^2 - 2(2+\varepsilon) \left\{ E(\varphi) - \int_0^t \int_\Omega u_t^2 \right\}. \quad (5.16)$$

Observe further that

$$2 \int_{\{|u|<K\}} u(t)g(u(t)) - 2(2+\varepsilon) \int_{\{|u|<K\}} G(u(t)) \geq 2|\Omega|(\mu - (2+\varepsilon)\nu). \quad (5.17)$$

Putting together (5.15), (5.16), and (5.17), we obtain the following inequality, for $0 \leq t < T(\varphi)$:

$$f'(t) \geq 2(2+\varepsilon) \int_0^t \int_\Omega u_t^2 + \varepsilon \int_\Omega |\nabla u(t)|^2 + 2|\Omega|(\mu - (2+\varepsilon)\nu) - 2(2+\varepsilon)E(\varphi). \quad (5.18)$$

In particular,

$$f'(t) \geq 2(2+\varepsilon) \int_0^t \int_\Omega u_t^2. \quad (5.19)$$

Now set

$$h(t) = \int_0^t f(s)\, ds.$$

Applying the Cauchy–Schwarz inequality, we obtain

$$h'(t) - h'(0) = f(t) - f(0) = 2 \int_0^t \int_\Omega u u_t$$
$$\leq 2 \left(\int_0^t \int_\Omega u^2 \right)^{1/2} \left(\int_0^t \int_\Omega u_t^2 \right)^{1/2} \leq 2 h(t)^{\frac{1}{2}} \left(\int_0^t \int_\Omega u_t^2 \right)^{1/2}.$$

From (5.19), it follows that

$$h(t) h''(t) \geq \left(1 + \frac{\varepsilon}{2} \right) (h'(t) - h'(0))^2. \quad (5.20)$$

For the sake of contradiction, assume that $T(\varphi) = \infty$. From (5.20), we have $(1 + \varepsilon/2)(h'(t) - h'(0))^2 \geq (1 + \varepsilon/4) h'(t)^2$ for $t \geq t_0$ large enough. It follows from (5.20) that

$$(h(t)^{-\varepsilon/4})'' \leq 0, \quad (5.21)$$

for $t \geq t_0$. But $h(t) > 0$ and $h(t)^{-\varepsilon/4} \to 0$ as $t \to \infty$. Thus there exists $t_1 \geq t_0$ such that $(h^{-\varepsilon/4})'(t_1) < 0$. Hence, by (5.21),

$$0 \leq h(t)^{-\varepsilon/4} \leq h(t_1)^{-\varepsilon/4} + (t - t_1)(h^{-\varepsilon/4})'(t_1),$$

for $t \geq t_1$; and so
$$t \leq t_1 - \frac{h(t_1)^{-\varepsilon/4}}{(h^{-\varepsilon/4})'(t_1)},$$
for $t \geq t_1$, which is absurd. □

Remark 5.4.7. The condition on g in Proposition 5.4.4 means that g is superlinear. Indeed, in the case in which there exists $x_0 > 0$ such that $G(x_0) > 0$, it means that $x^{-(2+\varepsilon)}G(x)$ is non-decreasing on $[x_0, \infty)$ (if $G(x_0) > 0$ for a certain $x_0 < 0$, then $x^{-(2+\varepsilon)}G(x)$ is non-increasing on $(-\infty, x_0]$). In this case, $G(x) \geq ax^{2+\varepsilon} - bx^2$, for $x \geq x_0$. Thus, if we take $\zeta \in X \cap H_0^1(\Omega)$ such that $\zeta \geq 0$, then $E(k\zeta) \to -\infty$ as $k \to \infty$. In particular, for $k > 0$ large enough the assumptions of Proposition 5.4.4 are fulfilled and so $T(k\zeta) < \infty$.

5.5. Application to a model case

We choose $g(x) = a|x|^\alpha x$, with $\alpha > 0$ and $a \neq 0$. We consider φX, and we denote by u the corresponding solution of (5.1)–(5.3). Then we have the following results.

If $a \leq 0$, then $T(\varphi) = \infty$, and u is bounded in X (Proposition 5.3.5).

If $a > 0$, then $T(\varphi) = \infty$ if $\|\varphi\|$ is small enough (Propositions 5.3.7 or 5.3.9). On the other hand, for some φ we have $T(\varphi) < \infty$ (Remarks 5.4.3 and 5.4.7), and in that case $\|u(t)\| \geq \delta(T(\varphi) - t)^{-\frac{1}{\alpha}}$ (Theorem 4.3.4).

Notes. We have studied the heat equation in the space $C_0(\Omega)$. It is also possible to study it in the spaces $C^{m,\alpha}(\Omega)$ (see Friedman [1], and Ladyzhenskaya, Solonikov and Ural'ceva [1]) and in the spaces $L^p(\Omega)$ (see Weissler [2]). In $L^p(\Omega)$, we observe certain singular phenomena, such as non-uniqueness (see Baras [1], Brezis, Peletier and Terman [1], and Haraux and Weissler [1]).

In some cases, the regularizing effect allows one to solve the Cauchy problem with singular initial data, such as measures (see Brezis and Friedman [1]). We can also consider more general non-linearities, depending on the derivatives of u, and more general elliptic operators than the Laplacian. See, for example, Friedman [1], Henry [1], and Ladyzhenskaya, Solonikov and Ural'ceva [1]. Some non-linearities with singularity at 0 have also been studied; see Aguirre and Escobedo [1]. Concerning systems, consult, for example, Dias and Haraux [1], Fife [1], and Smoller [1].

Various versions of the maximal principle for the heat equation can be found in Protter and Weinberger [1]. For the linear and non-linear regularizing effects, consult Friedman [1], Haraux and Kirane [1], Henry [1], Kirane and Tronel [1], and Ladyzhenskaya, Solonikov and Ural'ceva [1].

For more blow-up results, consult Fujita [1], Levine [1], and Payne and Sattinger [1]. The nature of blow-up is currently rather well known. See Baras and Cohen [1], Baras and Goldstein [1], Friedman and Giga [1], Friedman and McLeod [1], Giga and Kohn [1, 2], Mueller and Weissler [1], and Weissler [2, 3].

For the behaviour at infinity of solutions, consult Chapters 8 and 9, as well as, for example, Cazenave and Lions [2], Escobedo and Kavian [1, 2], Haraux [1], Henry [1], Kavian [2], Lions [1, 2], Weissler [4], and Esteban [1, 2].

6
The Klein–Gordon equation

6.1. Preliminaries

In this section, we give some technical tools that are essential in this chapter.

6.1.1. An abstract result

Let X be a Hilbert space, let A be a skew-adjoint operator in X, and let $(T(t))_{t\in\mathbb{R}}$ be the isometry semigroup generated by A. We have the following result.

Proposition 6.1.1. *Let $T > 0$, $x \in X$, and $f \in C([0,T], X)$. Let $u \in C([0,T], X)$ be given by*

$$u(t) = T(t)x + \int_0^t T(t-s)f(s)\,\mathrm{d}s,$$

for all $t \in [0,T]$. Then the function $t \mapsto \|u(t)\|^2$ belongs to $C^1([0,T])$ and

$$\frac{1}{2}\frac{\mathrm{d}}{\mathrm{d}t}\|u(t)\|^2 = \langle f(t), u(t) \rangle,$$

for all $t \in [0,T]$.

Proof. Suppose that $x \in D(A)$ and $f \in C^1([0,T], X)$. By Proposition 4.1.6, we have $u \in C([0,T], D(A)) \cap C^1([0,T], X)$ and

$$u'(t) = Au(t) + f(t),$$

for all $t \in [0,T]$. Thus,

$$\frac{1}{2}\frac{\mathrm{d}}{\mathrm{d}t}\|u(t)\|^2 = \langle u(t), u'(t) \rangle = \langle u(t), Au(t) + f(t) \rangle = \langle u(t), f(t) \rangle.$$

Hence

$$\|u(t)\|^2 = \|x\|^2 + \int_0^t \langle u(s), f(s) \rangle\,\mathrm{d}s. \tag{6.1}$$

In the general case, we approximate x and f by sequences $(x_n)_{n\geq 0} \subset D(A)$ and $(f_n)_{n\geq 0} \subset C^1([0,T], X)$, and then we pass to the limit to obtain (6.1). The result follows since u and f are continuous functions. □

6.1.2. Functionals on $H_0^1(\Omega)$

In this section, Ω is any open subset of $\mathbb{R}^N$. We consider a function $g \in C(\mathbb{R}, \mathbb{R})$ such that there exists $0 \le \alpha < \infty$ and $C < \infty$ so that

$$g(0) = 0; \tag{6.2}$$

$$|g(x) - g(y)| \le C(|x|^\alpha + |y|^\alpha)|x - y|, \quad \forall x, y \in \mathbb{R}. \tag{6.3}$$

In particular, we have $|g(x)| \le C|x|^{\alpha+1}$, and so, for all $p \ge \alpha + 1$, g defines an operator $F : L^p_{\mathrm{loc}}(\Omega) \to L^1_{\mathrm{loc}}(\Omega)$ by

$$F(u)(x) = g(u(x)),$$

for almost every $x \in \Omega$. When there is no risk of confusion, we still denote by g the operator F. Applying (6.2), (6.3) and Hölder's inequality, we easily obtain the following result.

Proposition 6.1.2. *Let $\alpha + 1 \le p \le \infty$. Then F is Lipschitz continuous from bounded subsets of $L^p(\Omega)$ to $L^{\frac{p}{\alpha+1}}(\Omega)$. More precisely, we have*

$$\|g(u) - g(v)\|_{L^{\frac{p}{\alpha+1}}} \le C(\|u\|_{L^p}^\alpha + \|v\|_{L^p}^\alpha)\|u - v\|_{L^p}, \tag{6.4}$$

for all $u, v \in L^p(\Omega)$.

For g as above, we define $G \in C(\mathbb{R}, \mathbb{R})$ by

$$G(x) = \int_0^x g(s)\,ds. \tag{6.5}$$

Then G verifies condition (6.3), with α replaced by $\alpha + 1$. Then, it follows from Proposition 6.1.2 that G allows us to define a functional V, Lipschitz continuous on bounded subsets of $L^{\alpha+2}(\Omega)$, by

$$V(u) = -\int G(u(x))\,dx, \quad \forall u \in L^{\alpha+2}(\Omega). \tag{6.6}$$

More precisely, we have the following.

Proposition 6.1.3. *V is a functional of class C^1 on $L^{\alpha+2}(\Omega)$. Its derivative (which is a continuous mapping $L^{\alpha+2}(\Omega) \to (L^{\alpha+2}(\Omega))' = L^{\frac{\alpha+2}{\alpha+1}}(\Omega)$) is given by*

$$V'(u) = -g(u), \tag{6.7}$$

for all $u \in L^{\alpha+2}(\Omega)$.

Proof. We have, for all $x, y \in \mathbb{R}$,

$$G(x+y) - G(x) - yg(x) = \int_x^{x+y} (g(\sigma) - g(x)) \, d\sigma;$$

and so, applying (6.3),

$$|G(x+y) - G(x) - yg(x)| \leq C'(|x|^\alpha + |y|^\alpha)|y|^2.$$

Applying Hölder's inequality, we deduce that, for all $u, v \in L^{\alpha+2}(\Omega)$,

$$\left| V(u+v) - V(u) + \langle v, g(u) \rangle_{L^{\alpha+2}, L^{\frac{\alpha+2}{\alpha+1}}} \right| = \left| V(u+v) - V(v) + \int vg(u) \, dx \right|$$
$$\leq C'(\|u\|_{L^{\alpha+2}}^\alpha + \|v\|_{L^{\alpha+2}}^\alpha)\|v\|_{L^{\alpha+2}}^2;$$

hence the result. □

We now assume that, instead of (6.3), g satisfies the following weaker condition:

$$|g(x) - g(y)| \leq C(1 + |x|^\alpha + |y|^\alpha)|x - y|, \tag{6.8}$$

for all $x, y \in \mathbb{R}$. In that case, we write

$$g = g_1 + g_2, \tag{6.9}$$

where g_2 verifies (6.3) and g_1 verifies (6.3) with $\alpha = 0$. For example, consider

$$g_1(x) = \begin{cases} g(x), & \text{if } |x| \leq 1; \\ g(1), & \text{if } x \geq 1; \\ g(-1), & \text{if } x \leq 1. \end{cases}$$

On the other hand, if $(N-2)p \leq 2N$, we have

$$H_0^1(\Omega) \hookrightarrow L^p(\Omega) \tag{6.10}$$

with dense embedding, and so

$$L^{p'}(\Omega) \hookrightarrow H^{-1}(\Omega). \tag{6.11}$$

The result is the following.

Proposition 6.1.4. *Let g satisfy (6.2) and (6.8), with $0 \leq \alpha \leq 4/(N-2)$. Then g is Lipschitz continuous from bounded subsets of $H_0^1(\Omega)$ to $H^{-1}(\Omega)$.*

Proof. By Proposition 6.1.2, g_1 is Lipschitz continuous from bounded subsets of $L^2(\Omega)$ to $L^2(\Omega)$, and so from bounded subsets of $H_0^1(\Omega)$ to $H^{-1}(\Omega)$. Invoking Proposition 6.1.2 again, g_2 is Lipschitz continuous from bounded subsets of $L^{\alpha+2}(\Omega)$ to $L^{\frac{\alpha+2}{\alpha+1}}(\Omega)$. Applying (6.10) and (6.11), with $p = \alpha+2$, we deduce that g_2 is Lipschitz continuous from bounded subsets of $H_0^1(\Omega)$ to $H^{-1}(\Omega)$; hence the result. □

Proposition 6.1.5. *Suppose that g satisfies the hypotheses of Proposition 6.1.4 with $(N-2)\alpha \leq 2$. Then g is Lipschitz continuous from bounded subsets of $H_0^1(\Omega)$ to $L^2(\Omega)$.*

Proof. The result is immediate for g_1. Set $\rho = 2(\alpha+2)$. Applying (6.3) and Hölder's inequality, it follows that

$$\|g_2(u) - g_2(v)\|_{L^2} \leq C(\|u\|_{L^\rho}^\alpha + \|v\|_{L^\rho}^\alpha)\|u-v\|_{L^{\alpha+2}},$$

for all $u, v \in H_0^1(\Omega)$. However, we have $(N-2)\rho \leq 2N$; hence the result, by (6.10). □

We now consider G and V defined by (6.5) and (6.6). We have the following result.

Proposition 6.1.6. *Suppose that g satisfies (6.2) and (6.8), with $\alpha \geq 0$ such that $(N-2)\alpha \leq 4$. Then V is a functional of class C^1 on $H_0^1(\Omega)$. Its derivative (which is a continuous mapping from $H_0^1(\Omega) \to (H_0^1(\Omega))' = H^{-1}(\Omega)$) is given by*

$$V'(u) = -g(u), \tag{6.12}$$

for all $u \in H_0^1(\Omega)$.

Proof. We apply Proposition 6.1.3 to g_1 and g_2, and we use embeddings (6.10) and (6.11). □

Corollary 6.1.7. *Suppose that g satisfies the hypotheses of Proposition 6.1.6, with $(N-2)\alpha \leq 2$. Let $T > 0$ and $u \in C([0,T], H_0^1(\Omega)) \cap C^1([0,T], L^2(\Omega))$. Then the mapping $t \mapsto V(u(t))$ is in $C^1([0,T])$, and we have*

$$\frac{\mathrm{d}}{\mathrm{d}t} V(u(t)) = -\int_\Omega g(u(t))u_t(t)\,\mathrm{d}x, \tag{6.13}$$

for all $t \in [0,T]$.

Proof. Suppose first that $u \in C^1([0,T], H_0^1(\Omega))$. Then, for all $t \in [0,T]$,

$$\frac{d}{dt}V(u(t)) = \langle V'(u(t)), u'(t)\rangle_{H^{-1}, H_0^1}$$
$$= -\langle g(u(t)), u'(t)\rangle_{H^{-1}, H_0^1} = -\int_\Omega g(u(t))u_t(t)\,dx.$$

It follows that

$$V(u(t)) = V(u(0)) - \int_0^t \int_\Omega g(u(s))u_t(s)\,ds. \tag{6.14}$$

By density, we deduce that (6.14) is still true when $u \in C([0,T], H_0^1(\Omega)) \cap C^1([0,T], L^2(\Omega))$; hence the result. $\square$

6.2. Local existence

Throughout this chapter, we follow the notation of §2.6.3, §2.6.4, and §3.5.2. In particular, Ω is any open subset of $\mathbb{R}^N$, $m > -\lambda$, $X = H_0^1(\Omega) \times L^2(\Omega)$ and $Y = L^2(\Omega) \times H^{-1}(\Omega)$. We consider a function $g \in C(\mathbb{R}, \mathbb{R})$ which satisfies (6.2) and (6.8) with $(N-2)\alpha \le 2$. Finally, we consider G and V defined by (6.5) and (6.6). We define the functional E on X and the mapping $F: X \to X$ by

$$E(u,v) = \frac{1}{2}\|(u,v)\|_X^2 + V(u)$$
$$= \frac{1}{2}\int_\Omega \{|\nabla u|^2 + m|u|^2 + |v|^2 - 2G(u)\}\,dx,$$
$$F(u,v) = (0, g(u)), \quad \text{for all } (u,v) \in H_0^1(\Omega) \times L^2(\Omega).$$

It is clear, from Proposition 6.1.5, that g defines a Lipschitz continuous mapping from $H_0^1(\Omega)$ to $L^2(\Omega)$, and so F is Lipschitz continuous on bounded subsets of X. Given $(\varphi, \psi) \in X$, we are looking for $T > 0$, and u a solution of

$$\begin{cases} u \in C([0,T], H_0^1(\Omega)) \cap C^1([0,T], L^2(\Omega)) \cap C^2([0,T], H^{-1}(\Omega)); & (6.15) \\ u_{tt} - \Delta u + mu = g(u), \quad \text{for all } t \in [0,T]; & (6.16) \\ u(0) = \varphi, \quad u_t(0) = \psi. & (6.17) \end{cases}$$

Applying Corollary 4.1.7 and Proposition 4.3.9, and arguing as in the proof of Proposition 3.5.11, we obtain the following result.

Lemma 6.2.1. Let $T > 0$ and $(\varphi, \psi) \in X$. Let $u \in C([0,T], H_0^1(\Omega)) \cap C^1([0,T], L^2(\Omega))$. Then u is solution of (6.15)–(6.17) if and only if $U = (u, u_t)$ is solution of

$$U(t) = \mathcal{T}(t)(\varphi,\psi) + \int_0^t \mathcal{T}(t-s)F(U(s))\,ds, \qquad (6.18)$$

for all $t \in [0,T]$. In addition, if $\triangle\varphi \in L^2(\Omega)$ and $\psi \in H_0^1(\Omega)$, then we have $u \in C^1([0,T], H_0^1(\Omega)) \cap C^2([0,T], L^2(\Omega))$ and $\triangle u \in C([0,T], L^2(\Omega))$.

Applying Theorem 4.3.4, we deduce a local existence result.

Theorem 6.2.2. *For all $(\varphi,\psi) \in X$, there exists a unique function u, defined on a maximal interval $[0, T(\varphi,\psi))$, which is a solution to (6.15)–(6.17) for all $T < T(\varphi,\psi)$. If, in addition, $T(\varphi,\psi) < \infty$, then $\|u(t)\|_{H^1} + \|u_t(t)\|_{L^2} \to \infty$ as $t \uparrow T(\varphi,\psi)$.*

Finally, we have conservation of energy.

Proposition 6.2.3. *Let $(\varphi,\psi) \in X$ and let u be the corresponding solution of (6.15)–(6.17). Then,*

$$E(u(t), u_t(t)) = E(\varphi,\psi), \qquad (6.19)$$

for all $t \in [0, T(\varphi,\psi))$.

Proof. We apply Proposition 6.1.1 and Corollary 6.1.7. It follows that

$$\frac{d}{dt}E(u(t), u_t(t)) = \langle F(u(t), u_t(t)), (u(t), u_t(t))\rangle_X - \int_\Omega g(u(t))u_t(t)\,dx = 0,$$

for all $t \in [0, T(\varphi,\psi))$; hence the result. □

Remark 6.2.4. Proposition 6.2.3 justifies the study of the Klein–Gordon equation in the space X. Indeed, the energy E is related to the X-norm and, as we will see in the next sections, the conservation of the energy (6.19) allows us, under certain hypotheses on g and (φ,ψ), to obtain estimates for the solution in X (and so global existence), or results about blow-up in finite time.

Remark 6.2.5. By using §4.4, we can solve problem (6.15)–(6.17) for $T < 0$ as well as for $T > 0$. Actually, note that u is a solution of (6.16) on $[-T, 0]$ with $u(0) = \varphi$ and $u_t(0) = \psi$ if and only if $v(t) = u(-t)$ is solution of (6.16) on $[0, T]$ with $v(0) = \varphi$ and $v_t(0) = -\psi$.

6.3. Global existence

As for the heat equation (§5.3), we will state two kinds of result according to the hypotheses on g: global existence of all solutions (i.e. independent of initial data), or global existence of solutions with small initial data.

Proposition 6.3.1. *Suppose that there exists $C < \infty$ such that $G(x) \leq C|x|^2$ for all $x \in \mathbb{R}$. Then, for all $(\varphi, \psi) \in X$, we have $T(\varphi, \psi) = \infty$.*

Proof. Set $f(t) = \|(u(t), u_t(t))\|_X^2$, for $t \in [0, T(\varphi, \psi))$. By (6.19), we have

$$f(t) \leq f(0) - 2\int G(\varphi) + 2\int G(u(t)) \leq f(0) - 2\int G(\varphi) + 2C\int u(t)^2, \quad (6.20)$$

for all $t \in [0, T(\varphi, \psi))$. On the other hand, we have

$$\begin{aligned}\|u(t)\|_{L^2}^2 &= \|\varphi\|_{L^2}^2 + \int_0^t \frac{d}{dt}\int_\Omega |u(s)|^2 = \|\varphi\|_{L^2}^2 + 2\int_0^t \int_\Omega u(s)u_t(s) \\ &\leq \|\varphi\|_{L^2}^2 + 2\int_0^t f(s).\end{aligned} \quad (6.21)$$

Applying (6.20), (6.21), and Gronwall's lemma, we obtain the result. □

Remark 6.3.2. If $2C < \lambda + m$ (λ being given by (2.2)), then for all $(\varphi, \psi) \in X$ the corresponding solution u of (6.15)–(6.17) satisfies $\sup_{t \geq 0} \|(u(t), u_t(t))\|_X < \infty$. Indeed, in this case we easily verify that $C \int u(t)^2 \leq (1-\varepsilon)f(t)$, with $\varepsilon > 0$, and it follows from (6.20) that $\varepsilon f(t) \leq C'$; hence the result.

Proposition 6.3.3. *Suppose that there exist $\mu < \lambda + m$ (λ being given by (2.2)) and $\beta > 0$ such that $2G(x) \leq \mu|x|^2$ for $|x| \leq \beta$. Then there exists $\delta, K > 0$ such that if $\|(\varphi, \psi)\|_X \leq \delta$, we have $T(\varphi, \psi) = \infty$ and the corresponding solution u of (6.15)–(6.17) satisfies $\sup_{t \geq 0} \|(u(t), u_t(t))\|_X \leq K\|(\varphi, \psi)\|_X$.*

Proof. The hypotheses on g imply that there exists a constant $C < \infty$ and $k > 2$ with $(N-2)k \leq 2N$, such that

$$2|G(x)| \leq C(|x|^2 + |x|^k),$$

for all $x \in \mathbb{R}$. By possibly taking larger C, we have

$$2G(x) \leq \mu|x|^2 + C|x|^k,$$

for all $x \in \mathbb{R}$. Sobolev's inequalities show that there exists a constant, which we will still denote by C, such that

$$2\int G(u) \leq \mu \int u^2 + C\|u\|_{H^1}^k,$$

for all $u \in H_0^1(\Omega)$. Furthermore, since $\mu < \lambda + m$, we easily verify that there exists $\nu > 0$ such that

$$\mu \int u^2 \leq (1-\nu)\|u\|_{H^1}^2,$$

for all $u \in H_0^1(\Omega)$, and consequently

$$2\int G(u) \leq (1-\nu)\|u\|_{H^1}^2 + C\|u\|_{H^1}^k, \qquad (6.22)$$

for all $u \in H_0^1(\Omega)$. Let $(\varphi, \psi) \in X$, with $\|(\varphi, \psi)\|_X \leq 1$, and let u be the corresponding solution to (6.15)–(6.17). Using the notation of the proof of Proposition 6.3.1, we deduce from (6.20) and (6.22) that, for all $t \in [0, T(\varphi, \psi))$,

$$\nu f(t) \leq f(0) - 2\int G(\varphi) + Cf(t)^{\frac{k}{2}}. \qquad (6.23)$$

Observe that

$$\left|\int G(\varphi)\right| \leq C(\|(\varphi, \psi)\|_X^2 + \|(\varphi, \psi)\|_X^k) \leq C(f(0)^2 + f(0)^k);$$

and so, since $f(0) \leq 1$, there exists $M \in [1, \infty)$ such that

$$f(0) - 2\int G(\varphi) \leq \nu M f(0). \qquad (6.24)$$

Set $k/2 = 1 + \varepsilon$ ($\varepsilon > 0$) and

$$\theta(x) = \frac{C}{\nu}x^{1+\varepsilon} - x,$$

for $x \geq 0$, and set $-\chi = \min \theta < 0$. For all $a \in (0, \chi)$ there exists $0 < x_a < y_a$ such that

$$\theta(x_a) + a = \theta(y_a) + a = 0.$$

In addition, we have $a < x_a < a(1+\varepsilon)/\varepsilon$ (see Figure 6.1).

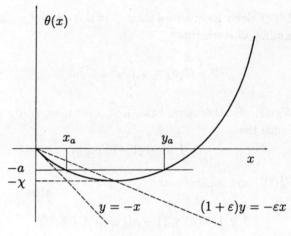

Fig. 6.1

By (6.23) and (6.24), we have

$$\theta(f(t)) + Mf(0) \geq 0,$$

for all $t \in [0, T(\varphi, \psi))$. Consequently, if we suppose that $Mf(0) < \chi$, we have

$$f(t) \in [0, x_{Mf(0)}) \cup (y_{Mf(0)}, \infty),$$

and since f is continuous and $f(0) < x_{Mf(0)}$, we deduce that

$$f(t) \leq x_{Mf(0)} \leq \frac{1+\varepsilon}{\varepsilon} Mf(0),$$

for all $t \in [0, T(\varphi, \psi))$. The result follows, with $\delta = \min\{1, \sqrt{\chi/M}\}$ and $K = \sqrt{(1+\varepsilon)/\varepsilon M}$. □

Remark 6.3.4. If $\lambda = 0$ (for example, if $\Omega = \mathbb{R}^N$), we may apply Proposition 6.3.3 only if $G(x) < 0$ for $|x|$ small. Then, we can replace the hypotheses $G(x) \leq \mu|x|^2$ by $G(x) \leq \mu|x|^{2+\varepsilon}$, for $|x|$ small ($\varepsilon > 0$), the proof being the same.

Remark 6.3.5. When Ω is bounded, we may assume that the conditions on G involved in the statements of Proposition 6.3.1 and Remark 6.3.2 hold only for $|x|$ large. The proof is the same; see Proposition 6.4.4.

6.4. Blow-up in finite time

The main result of this section is the following.

Proposition 6.4.1. *Suppose that there exists $\varepsilon > 0$ such that*
$$xg(x) \geq (2+\varepsilon)G(x),$$
for all $x \in \mathbb{R}$. Then, if $(\varphi, \psi) \in X$ and $E(\varphi, \psi) < 0$, we have $T(\varphi, \psi) < \infty$.

The proof of Proposition 6.4.1 requires the following lemmas.

Lemma 6.4.2. *Let $T > 0$ and let $u \in C([0,T], H_0^1(\Omega)) \cap C^1([0,T], L^2(\Omega)) \cap C^2([0,T], H^{-1}(\Omega))$. Then the function $t \mapsto \int u(t)^2$ belongs to $C^2([0,T])$ and we have*
$$\frac{d^2}{dt^2} \int_\Omega u(t)^2 \, dx = 2 \int_\Omega u_t(t)^2 \, dx + 2\langle u(t), u_{tt}(t) \rangle_{H_0^1, H^{-1}},$$
for all $t \in [0,T]$.

Proof. Set
$$f(t) = \int_\Omega u(t)^2, \quad \forall t \in [0,T],$$
and assume first that $u \in C^2([0,T], H_0^1(\Omega))$. Then
$$f''(t) = 2\int_\Omega u_t(t)^2 + 2\int_\Omega u_t(t)u_{tt}(t) = 2\int_\Omega u_t(t)^2 + 2\langle u(t), u_{tt}(t) \rangle_{H_0^1, H^{-1}},$$
for all $t \in [0,T]$. It follows that
$$f'(t) = f'(0) + 2\int_0^t \left\{ \int_\Omega u_t(s)^2 + \langle u(s), u_{tt}(s) \rangle_{H_0^1, H^{-1}} \right\} ds, \qquad (6.25)$$
for all $t \in [0,T]$. By density, we then show that (6.25) still holds for $u \in C([0,T], H_0^1(\Omega)) \cap C^1([0,T], L^2(\Omega)) \cap C^2([0,T], H^{-1}(\Omega))$; hence the result. □

Lemma 6.4.3. *Let $T > 0$ and let $u \in C([0,T], H_0^1(\Omega)) \cap C^1([0,T], L^2(\Omega)) \cap C^2([0,T], H^{-1}(\Omega))$ be the solution of (6.16). Set*
$$f(t) = \int_\Omega u(t)^2,$$
for all $t \in [0,T]$. Then
$$f''(t) = 2\int_\Omega u_t(t)^2 - 2\int_\Omega |\nabla u(t)|^2 - 2m\int_\Omega u(t)^2 + 2\int_\Omega u(t)g(u(t)),$$
for all $t \in [0,T]$.

Proof. We apply Lemma 6.4.2 and (6.16). It follows that

$$f''(t) = 2\int_\Omega u_t(t)2 + 2\langle u(t), \Delta u(t) - mu(t) + g(u(t))\rangle_{H_0^1, H^{-1}},$$

for all $t \in [0,T]$. It suffices to see that, for all $w \in H_0^1(\Omega)$, we have

$$\langle w, \Delta w\rangle_{H_0^1, H^{-1}} = -\int_\Omega |\nabla w|^2.$$

This is a consequence of Lemma 2.6.2 if $\Delta w \in L^2(\Omega)$, and follows by density otherwise. □

Proof of Proposition 6.4.1. Let $(\varphi, \psi) \in X$ be such that $E(\varphi, \psi) < 0$, and set

$$f(t) = \int_\Omega u(t)^2,$$

for all $t \in [0, T(\varphi, \psi))$. By Lemma 6.4.3, we have

$$f''(t) = 2\int_\Omega u_t(t)^2 - 2\int_\Omega |\nabla u(t)|^2 - 2m\int_\Omega u(t)^2 + 2\int_\Omega u(t)g(u(t))$$

$$\geq 2\int_\Omega u_t(t)^2 - 2\int_\Omega |\nabla u(t)|^2 - 2m\int_\Omega u(t)^2 + 2(2+\varepsilon)\int_\Omega G(u(t)),$$

for all $t \in [0, T(\varphi, \psi))$. Applying Proposition 6.2.3, we deduce that

$$f''(t) \geq \varepsilon\left\{\int_\Omega |\nabla u(t)|^2 + m\int_\Omega u(t)^2\right\} + (4+\varepsilon)\int_\Omega u_t(t)^2 - 2(2+\varepsilon)E(\varphi, \psi), \quad (6.26)$$

for all $t \in [0, T(\varphi, \psi))$. For the sake of contradiction, assume that $T(\varphi, \psi) = \infty$. In particular, we deduce from (6.26) that

$$f''(t) \geq -2(2+\varepsilon)E(\varphi, \psi) > 0, \quad \forall t \geq 0.$$

It follows that $f(t) \to \infty$ as $t \to \infty$. On the other hand, applying (6.26) and the Cauchy–Schwarz inequality, it follows that

$$f(t)f''(t) \geq (4+\varepsilon)\int_\Omega u_t(t)^2\int_\Omega u(t)^2 \geq (4+\varepsilon)\left(\int_\Omega u_t(t)u(t)\right)^2 \geq \left(1+\frac{\varepsilon}{4}\right)f'(t)^2,$$

and then

$$f(t)^{-\frac{\varepsilon}{4}} \leq 0, \quad \forall t \geq 0.$$

We conclude as for Proposition 5.4.4 (inequality (5.21) and below). □

In the case in which Ω is bounded, we can weaken the hypotheses of Proposition 6.4.1.

Proposition 6.4.4. *Suppose that Ω is bounded and that there exist $K < \infty$ and $\varepsilon > 0$ such that*
$$xg(x) \geq (2+\varepsilon)G(x),$$
for $|x| \geq K$. Set $\mu = \min_{|x| \leq K} xg(x)$ and $\nu = \max_{|x| \leq K} G(x)$. Then, if $(\varphi, \psi) \in X$ satisfies
$$(2+\varepsilon)E(\varphi, \psi) < |\Omega|(\mu - (2+\varepsilon)\nu),$$
we have $T(\varphi, \psi) < \infty$.

Proof. We argue as in Proposition 6.4.1, using
$$\int_\Omega u(t)g(u(t)) = \int_{\{|u| \leq K\}} u(t)g(u(t)) + \int_{\{|u| > K\}} u(t)g(u(t))$$
$$\geq m|\Omega| + (2+\varepsilon)\int_{\{|u| > K\}} G(u(t))$$
$$\geq m|\Omega| + (2+\varepsilon)\int G(u(t)) - (2+\varepsilon)\int_{\{|u| \leq K\}} G(u(t))$$
$$\geq (m - (2+\varepsilon)\nu)|\Omega| + (2+\varepsilon)\int G(u(t)),$$
for all $t \in [0, T(\varphi, \psi))$. □

Remark 6.4.5. For some comments on the hypotheses of Proposition 6.4.1 and 6.19, see Remark 5.4.7. In particular, if there exists $x_0 \geq K$ such that $G(x_0) > 0$, and if we write $(\zeta, \eta) \in X$ with $\zeta \geq 0$ a.e. on Ω, we have $E(k\zeta, k\eta) \to -\infty$ as $k \to \infty$. In particular, for k sufficiently large, $(k\zeta, k\eta)$ satisfies the conditions of Propositions 6.16 or 6.19, and so $T(k\zeta, k\eta) < \infty$.

6.5. Application to a model case

We choose $g(x) = a|x|^\alpha x$, with $a \neq 0$, $\alpha > 0$, and $(N-2)\alpha \leq 2$. We consider $(\varphi, \psi) \in X$ and we denote by u the corresponding solution of (6.15)–(6.17). Then we have the following results.

If $a < 0$, then $T(\varphi, \psi) = \infty$ and (u, u_t) is bounded in $H_0^1(\Omega) \times L^2(\Omega)$ (Remark 6.3.2).

If $a > 0$, then $T(\varphi, \psi) = \infty$ if $\|(\varphi, \psi)\|_X$ is small enough (Proposition 6.3.3). In addition, for some $(\varphi, \psi) \in X$, we have $T(\varphi, \psi) < \infty$ (Remark 6.4.5), and in that case $\|(u(t), u_t(t))\|_X \geq \delta(T(\varphi, \psi) - t)^{-\frac{1}{\alpha}}$ (Theorem 4.3.4).

Notes. For more about local and global existence, in the framework of Chapter 6, consult Browder [1] and Heinz and Von Wahl [2], and for more about blow-up phenomenon, see Levine [2, 3], J. B. Keller [1], and Glassey [1, 3]. In the general case ($\Omega \neq \mathbb{R}^N$) the behaviour at infinity of solutions is well known only in the dissipative case. See §9.4 and, for example, Haraux [1, 2]. In the conservative case, we only have some partial results, often limited to dimension 1. See Brezis, Coron, and Nirenberg [1], Cabannes and Haraux [1, 2], Cazenave and Haraux [2, 3], Cazenave, Haraux, Vazquez, and Weissler [1], Friedlander [1], C. Keller [1], Payne and Sattinger [1], and Rabinowitz [1, 2]. On conservation laws, see Serre [1].

For $\Omega = \mathbb{R}^N$, there exist estimates of the same kind as in §7.3; see Brenner [1, 2], Ginibre and Velo [6, 9], and Marshall, Strauss, and Wainger [1]. These estimates allow us, for the local existence, to replace in (6.8) the condition $(N-2)\alpha \leq 2$ by the weaker condition $(N-2)\alpha \leq 4$. See Ginibre and Velo [8, 10] and Jörgens [1]. If condition (6.8) is not satisfied, we only know how to build solutions in the case $xg(x) \leq 0$, but we do not know whether uniqueness holds. See Strauss [1,2], as well as the very interesting numerical study of Strauss and Vazquez [1].

Again for $\Omega = \mathbb{R}^N$, we know how to investigate the dispersive properties of the linear equation to show the global existence for small initial conditions, with non-linearities that depend on derivatives of u (Klainerman and Ponce [1]). If the order of the non-linearity at 0 is not sufficiently high, blow-up in finite time may occur for arbitrarily small initial data. See Balabane [1, 2], Sideris [2, 3], Hanouzet and Joly [1], and John [1–3]. For some non-linearities, there exist solutions of the form $u(t, x) = e^{i\omega t}\varphi(x)$. These solutions are called stationary states. See, for example, Berestycki, Gallouet, and Kavian [1], Berestycki and Lions [1], Berestycki, Lions, and Peletier [1], and Jones and Küpper [1]. The behaviour at infinity of solutions is rather well known in the repulsive case, in which the solutions behave asymptotically as the solutions of the linear equation. See Brenner [1, 2], Ginibre and Velo [8, 10], Morawetz and Strauss [1], Reed and Simon [1], and Sideris [1]. In the attractive case, we know mainly how to study the stability of certain stationary states. See Berestycki and Cazenave [1], Blanchard, Stubbe, and Vazquez [1], Cazenave [3], Cazenave and Lions [1], Grillakis, Shatah, and Strauss [1], C. Keller [1], Payne and Sattinger [1], and Shatah and Strauss [1].

7
The Schrödinger equation

7.1. Preliminaries

Throughout this chapter, we use the notation of §2.6.5 and §3.5.3. In particular, $X = H^{-1}(\Omega) = H^{-1}(\Omega, \mathbb{C})$ and $Y = L^2(\Omega) = L^2(\Omega, \mathbb{C})$. The isometry groups generated by A and B are both denoted by $(\mathcal{T}(t))_{t \in \mathbb{R}}$ (see Lemma 3.5.12). Given $g : H_0^1(\Omega) \to H^{-1}(\Omega)$, Lipschitz continuous on bounded subsets and $\varphi \in H_0^1(\Omega)$, we are looking for $T > 0$, and u a solution of the following problem:

$$\begin{cases} u \in C([0,T], H_0^1(\Omega)) \cap C^1([0,T], H^{-1}(\Omega)); & (7.1) \\ iu_t + \Delta u + g(u) = 0, \quad \forall t \in [0,T]; & (7.2) \\ u(0) = \varphi. & (7.3) \end{cases}$$

Applying Lemma 4.1.1 and Corollary 4.1.8, we obtain the following result.

Lemma 7.1.1. *Let $T > 0$, $\varphi \in H_0^1(\Omega)$ and let $u \in C([0,T], H_0^1(\Omega))$. Then u is a solution of (7.1)–(7.3), if and only if u is solution of*

$$u(t) = \mathcal{T}(t)\varphi + i \int_0^t \mathcal{T}(t-s) g(u(s)) \, \mathrm{d}s, \qquad (7.4)$$

for all $t \in [0,T]$.

On the other hand, applying Proposition 4.1.9, we obtain a sharpened version of Lemma 7.1.1, which will also be used in what follows.

Lemma 7.1.2. *Let $T > 0$, $\varphi \in H_0^1(\Omega)$ and let $u \in L^\infty((0,T), H_0^1(\Omega))$. Then u is a solution of*

$$\begin{cases} u \in L^\infty((0,T), H_0^1(\Omega)) \cap W^{1,\infty}((0,T), H^{-1}(\Omega)); & (7.5) \\ iu_t + \Delta u + g(u) = 0, \quad \text{a.e. } t \in [0,T]; & (7.6) \\ u(0) = \varphi, & (7.3) \end{cases}$$

if and only if u is solution of (7.4).

Remark 7.1.3. If (7.1)–(7.3) can be solved for $T > 0$, then in general it can also be solved for $T < 0$ (see §4.4). If g satisfies certain symmetry properties, it is especially clear. Indeed, if we suppose that $g(\overline{u}) = \overline{g(u)}$, then v given by $v(t) = \overline{u(-t)}$ is a solution of (7.2) on $[0,T]$ if and only if u is a solution of (7.2) on $[-T,0]$.

7.2. A general result

Assume that $g : \mathbb{R}_+ \to \mathbb{R}$ is (globally) Lipschitz continuous and that $g(0) = 0$. We extend g to the complex plane by setting

$$g(z) = \frac{z}{|z|} g(|z|), \tag{7.7}$$

for all $z \in \mathbb{C}$, $z \neq 0$. We also define the function G by

$$G(z) = \int_0^{|z|} g(s)\,ds, \tag{7.8}$$

for all $z \in \mathbb{C}$. Then, g defines a mapping $L^2(\Omega) \to L^2(\Omega)$, which we still denote by g^*, by

$$g(v)(x) = g(v(x)), \tag{7.9}$$

for all $v \in L^2(\Omega)$ and for almost all $x \in \Omega$. In addition, g is Lipschitz on $L^2(\Omega)$ (see §6.1.2). Furthermore, if we set

$$V(v) = -\int_\Omega G(v(x))\,dx, \quad \forall v \in L^2(\Omega), \tag{7.10}$$

then $V \in C^1(L^2(\Omega), \mathbb{R})$ and

$$V'(v) = -g(v), \quad \forall v \in L^2(\Omega). \tag{7.11}$$

Finally, we define the functional $E \in C^1(H_0^1(\Omega), \mathbb{R})$ by

$$E(v) = \frac{1}{2}\int_\Omega |\nabla v|^2\,dx + V(v), \tag{7.12}$$

for all $v \in H_0^1(\Omega)$. We have the following result.

Theorem 7.2.1. *For all $\varphi \in L^2(\Omega)$, there exists a unique function $u \in C([0,\infty), L^2(\Omega))$ which is a solution to (7.4) for all $T < \infty$; furthermore,*

$$\|u(t)\|_{L^2} = \|\varphi\|_{L^2}, \tag{7.13}$$

---

* This is a common abuse of notation, g denotes both a mapping $\mathbb{C} \to \mathbb{C}$ and a mapping $L^2 \to L^2$. However, the context makes it clear which is used.

for all $t \geq 0$. If we assume further that $\varphi \in H_0^1(\Omega)$, then $u \in C([0,\infty), H_0^1(\Omega)) \cap C^1([0,\infty), H^{-1}(\Omega))$ and
$$E(u(t)) = E(\varphi), \qquad (7.14)$$
for all $t \geq 0$. If in addition $\Delta \varphi \in L^2(\Omega)$, then $\Delta u \in C([0,\infty), L^2(\Omega))$ and $u \in C^1([0,\infty), L^2(\Omega))$.

The proof makes use of the following lemma.

Lemma 7.2.2. Let $T > 0$ and let $u \in C([0,T], D(B)) \cap C^1([0,T], L^2(\Omega))$. Then the function $t \mapsto \|\nabla u(t)\|_{L^2}^2$ belongs to $C^1([0,T])$ and we have
$$\frac{d}{dt} \|u(t)\|_{L^2}^2 = 2\langle -\Delta u(t), u_t(t)\rangle,$$
for all $t \in [0,T]$.

Proof. The result is clear if $u \in C^1([0,T], D(B))$, and is obtained by density in the general case (see, for example, the proof of Proposition 6.1.1). □

Proof of Theorem 7.2.1. We proceed in seven steps.

Step 1. $\varphi \in L^2(\Omega)$. The global existence in $L^2(\Omega)$ is a consequence of Theorem 4.3.4, since $K(M)$ is bounded.

Step 2. $\varphi \in D(B)$. If $\varphi \in D(B)$, the regularity is a consequence of Proposition 4.3.9. In that case, u is solution of (7.2) for all $T < \infty$, and (7.2) is satisfied in $L^2(\Omega)$.

Step 3. The conservation law (7.13). If $\varphi \in D(B)$, and taking the scalar product (in $L^2(\Omega)$) of (7.2) with u, we obtain
$$\frac{1}{2}\frac{d}{dt}\|u(t)\|_{L^2}^2 = \langle u_t(t), u(t)\rangle = \langle i\Delta u(t), u(t)\rangle + \langle ig(u(t)), u(t)\rangle = 0. \qquad (7.15)$$

Indeed, for all $w \in D(B)$, we have (see Lemma 2.6.2)
$$\langle i\Delta u(t), u(t)\rangle - \operatorname{Re}\int_\Omega i\Delta w \overline{w} = -\operatorname{Re}\int_\Omega i|\nabla w|^2 = 0;$$
and (applying (7.8))
$$\langle ig(u(t)), u(t)\rangle = \operatorname{Re}\int_\Omega ig(w)\overline{w} = \operatorname{Re}\int_\Omega ig(|w|)|w| = 0.$$

Then (7.13) follows from (7.15). In the general case $\varphi \in L^2(\Omega)$, we obtain (7.13) by density, applying Proposition 4.3.7.

Step 4. The conservation law (7.14) for $\varphi \in D(B)$. Taking the scalar product (in $L^2(\Omega)$) of (7.2) with u_t, we obtain

$$0 = \operatorname{Re} \int_\Omega i|u_t|^2 = \langle iu_t, u_t\rangle = -\langle \Delta u, u_t\rangle - \langle g(u), u_t\rangle,$$

for all $t \in [0, \infty)$. Applying Lemma 7.2.2 and Corollary 6.1.7, we deduce that $(d/dt)E(u(t)) = 0$; hence (7.14).

Step 5. If $\varphi \in H_0^1(\Omega)$, then u is weakly continuous as a map from $[0, \infty)$ to $H_0^1(\Omega)$ and we have

$$E(u(t)) \leq E(\varphi), \tag{7.16}$$

for all $t \in [0, \infty)$. Indeed, consider a sequence $(\varphi_n)_{n\geq 0} \subset D(B)$ such that $\varphi_n \longrightarrow \varphi$ in $H_0^1(\Omega)$ as $n \to \infty$, and let u_n be the corresponding solutions of (7.2). Let $T > 0$. Since g is Lipschitz continuous on $L^2(\Omega)$, it follows from (7.13) and (7.14) that u_n is bounded in $L^\infty((0, T), H_0^1(\Omega))$. By Proposition 4.3.7, we also have

$$u_n \longrightarrow u \quad \text{in } C([0,T], L^2(\Omega)) \text{ as } n \to \infty. \tag{7.17}$$

In particular, for all $t \in [0, T]$, we have $u_n(t) \longrightarrow u(t)$ in $L^2(\Omega)$ as $n \to \infty$, and $\|u_n(t)\|_{H^1}$ is bounded. Therefore,

$$u_n(t) \rightharpoonup u(t) \quad \text{in } H_0^1(\Omega) \text{ as } n \to \infty. \tag{7.18}$$

Applying Proposition 1.4.24, we deduce that $u \in L^\infty((0, T), H_0^1(\Omega))$. Since $u \in C([0, T], L^2(\Omega))$, it follows that u is weakly continuous from $[0, T]$ to $H_0^1(\Omega)$. Applying (7.17), (7.18), and the weak lower semicontinuity of the norm in $H_0^1(\Omega)$, we deduce (7.16) from (7.14). We conclude observing that T is arbitrary.

Step 6. The conservation law (7.14) for $\varphi \in H_0^1(\Omega)$. For all $s \in [0, \infty)$, we set

$$v(t) = \overline{u(s-t)}, \quad \forall t \in [0, s].$$

On checking, it is immediate that v is solution of (7.2) on $[0, s]$. Then, we may apply (7.16). It follows that

$$E(\varphi) \leq E(u(s)). \tag{7.19}$$

By putting together (7.18) and (7.19), we obtain (7.15).

Step 7. If $\varphi \in H_0^1(\Omega)$, we have $u \in C([0, \infty), H_0^1(\Omega)) \cap C^1([0, \infty), H^{-1}(\Omega))$. Indeed, since g is continuous on $L^2(\Omega)$, we deduce from (7.14) that the function $t \mapsto \|u(t)\|_{H^1}$ belongs to $C([0, \infty))$. Since u is weakly continuous from $[0, \infty)$ to $H_0^1(\Omega)$, we then have $u \in C([0, \infty), H_0^1(\Omega))$; and so by Corollary 4.1.8, $u \in C^1([0, \infty), H^{-1}(\Omega))$. This completes the proof. □

Remark 7.2.3. Theorem 7.2.1 applies only if the non-linearity g is mild. Indeed, if we consider $g(u) = |u|^\alpha u$ ($\alpha \geq 0$), we may use it only if $\alpha = 0$. However, Theorem 7.2.1 will be useful to prove a more general result for $\Omega = \mathbb{R}^N$ (§7.4). It is then necessary to specify the dispersive properties of the Schrödinger equation in $\mathbb{R}^N$. These properties are described in the next section.

7.3. The linear Schrödinger equation in $\mathbb{R}^N$

We suppose that $\Omega = \mathbb{R}^N$, and we consider $T > 0$. We are going to apply the results of §3.5.4 to give estimates for the solutions of the inhomogeneous Schrödinger equation

$$\begin{cases} iu_t + \Delta u + f = 0; \\ u(0) = \varphi. \end{cases}$$

To do this, we define the operators Φ, Ψ, and Θ_t (for $t \in [0,T]$) by

$$\Phi_f(t) = \int_0^t T(t-s)f(s)\,ds, \quad \forall t \in [0,T];$$

$$\Psi_f(s) = \int_s^T T(s-t)f(t)\,dt, \quad \forall s \in [0,T];$$

$$\Theta_{t,f}(s) = \int_0^t T(s-\sigma)f(\sigma)\,d\sigma, \quad \forall s \in [0,T];$$

for all $f \in L^1((0,T), H^{-1}(\mathbb{R}^N))$. We easily verify (see Lemma 4.1.5) that Φ, Ψ, and Θ_t are continuous from $L^1((0,T), H^{-1}(\mathbb{R}^N))$ to $C([0,T], H^{-1}(\mathbb{R}^N))$, and from $L^1((0,T), H^1(\mathbb{R}^N))$ to $C([0,T], H^1(\mathbb{R}^N))$.

Definition 7.3.1. We say that a pair (q,r) of positive numbers is admissible if the following properties hold:

(i) $2 \leq r < 2N/(N-2)$ ($2 \leq r < \infty$ if $N = 2$, $2 \leq r \leq \infty$ if $N = 1$);

(ii) $2/q = N(1/2 - 1/r)$.

Observe that if (q,r) is an admissible pair then we have $q \in (2,\infty]$ ($q \in [4,\infty]$ if $N = 1$). The pair $(\infty, 2)$ is always admissible.

Remark 7.3.2. If (q,r) is an admissible pair, then we have in particular $H^1(\mathbb{R}^N) \hookrightarrow L^r(\mathbb{R}^N)$ with dense embedding and so $L^{r'}(\mathbb{R}^N) \hookrightarrow H^{-1}(\mathbb{R}^N)$. It follows that $C([0,T], H^1(\mathbb{R}^N)) \hookrightarrow L^q((0,T), L^r(\mathbb{R}^N))$ and $L^{q'}((0,T), L^{r'}(\mathbb{R}^N)) \hookrightarrow L^1((0,T), H^{-1}(\mathbb{R}^N))$. In particular, the operators Φ, Ψ, and Θ_t are continuous from $L^{q'}((0,T), L^{r'}(\mathbb{R}^N))$ to $C([0,T], H^{-1}(\mathbb{R}^N))$.

Remark 7.3.3. If (q,r) is an admissible pair, and taking suitable test functions, we easily verify that for all $u \in L^q((0,T), L^r(\mathbb{R}^N))$, we have

$$\|u\|_{L^q((0,T),L^r)} = \sup\left\{ \left| \int_0^T \operatorname{Re} \int_{\mathbb{R}^N} u\overline{\varphi}\,dx\,dt \right|;\right.$$
$$\left. \varphi \in L^{q'}((0,T), L^{r'}(\mathbb{R}^N)), \|\varphi\|_{L^{q'}((0,T),L^{r'})} \le 1 \right\}.$$

By truncation and regularization, we also have

$$\|u\|_{L^q((0,T),L^r)} = \sup\left\{ \left| \int_0^T \operatorname{Re} \int_{\mathbb{R}^N} u\overline{\varphi}\,dx\,dt \right|;\right.$$
$$\left. \varphi \in L^{q'}((0,T), L^{r'}(\mathbb{R}^N)) \cap C([0,T], H^1(\mathbb{R}^N)), \|\varphi\|_{L^{q'}((0,T),L^{r'})} \le 1 \right\}.$$

The main result of this section is the following.

Proposition 7.3.4. Let (γ, ρ) be an admissible pair according to Definition 7.3.1 and let $f \in L^{\gamma'}((0,T), L^{\rho'}(\mathbb{R}^N))$. Then $\Phi_f \in C([0,T], L^2(\mathbb{R}^N))$. In addition, for any admissible pair (q,r), we have $\Phi_f \in L^q((0,T), L^r(\mathbb{R}^N))$. Finally, there exists a constant C, depending only on γ and q, such that

$$\|\Phi_f\|_{L^q((0,T),L^r)} \le C(\gamma, q) \|f\|_{L^{\gamma'}((0,T),L^{\rho'})}, \qquad (7.20)$$

for all $f \in L^{\gamma'}((0,T), L^{\rho'}(\mathbb{R}^N))$.

Proof. The proof proceeds in six steps.

Step 1. For all admissible pairs (q,r), the operator $\Phi \in \mathcal{L}(L^{q'}((0,T), L^{r'}(\mathbb{R}^N)), L^q((0,T), L^r(\mathbb{R}^N)))$, with norm depending only on q. By density, we need only consider the case in which $f \in C([0,T], L^{r'}(\mathbb{R}^N))$. In this case, it follows from Proposition 3.5.14 that $\Phi_f \in C([0,T], L^r(\mathbb{R}^N))$, and that for $t \in [0,T]$ we have

$$\|\Phi_f(t)\|_{L^r} \le \int_0^t |t-s|^{-n(\frac{1}{2}-\frac{1}{r})} \|f(s)\|_{L^{r'}}\,ds \le \int_0^T |t-s|^{\frac{-2}{q}} \|f(s)\|_{L^{r'}}\,ds.$$

The classical estimates on Riesz' potentials (see Stein [1], Thm. 1, p. 119) then give

$$\|\Phi_f\|_{L^q((0,T),L^r)} \le C\|f\|_{L^{q'}((0,T),L^{r'})},$$

where C depends only on q; hence the result.

Step 2. Similarly, Ψ and Θ_t are continuous from $L^{q'}((0,T), L^{r'}(\mathbb{R}^N))$ to $L^q((0,T), L^r(\mathbb{R}^N))$ with norms depending only on q.

Step 3. The operator $\Phi \in \mathcal{L}(L^{q'}((0,T), L^{r'}(\mathbb{R}^N)), C([0,T], L^2(\mathbb{R}^N)))$, for all admissible pairs (q,r), with norm depending only on q. By density, we need only consider the case in which $f \in C([0,T], L^{r'}(\mathbb{R}^N))$, and using a regularizing sequence, we may assume further that $f \in C([0,T], H^1(\mathbb{R}^N))$. In that case, we have $\Phi_f \in C([0,T], H^1(\mathbb{R}^N))$. Applying Corollary 3.2.6, and denoting by $\langle \cdot, \cdot \rangle$ the scalar product in $L^2(\mathbb{R}^N)$, we obtain, for all $t \in [0,T]$,

$$\|\Phi_f(t)\|_{L^2}^2 = \langle \int_0^t \mathcal{T}(t-s)f(s)\,ds, \int_0^t \mathcal{T}(t-\sigma)f(\sigma)\,d\sigma \rangle$$

$$= \int_0^t \int_0^t \langle \mathcal{T}(t-s)f(s), \mathcal{T}(t-\sigma)f(\sigma) \rangle\,d\sigma\,ds$$

$$= \int_0^t \int_0^t \langle f(s), \mathcal{T}(s-\sigma)f(\sigma) \rangle\,d\sigma\,ds = \int_0^t \langle f(s), \Theta_{t,f}(s) \rangle\,ds.$$

Applying Hölder's inequality, first in space and then in time, and using Step 2, it follows that

$$\|\Phi_f(t)\|_{L^2}^2 \le \|f\|_{L^{q'}((0,T),L^{r'})} \|\Theta_{t,f}\|_{L^q((0,T),L^r)} \le C(q)\|f\|_{L^{q'}((0,T),L^{r'})}^2;$$

hence the result, since t is arbitrary.

Step 4. Similarly, Ψ and Θ_t are continuous from $L^{q'}((0,T), L^{r'}(\mathbb{R}^N))$ to $C([0,T], L^2(\mathbb{R}^N))$, with norms depending only on q.

Step 5. The operator $\Phi \in \mathcal{L}(L^1((0,T), L^2(\mathbb{R}^N)), L^q((0,T), L^r(\mathbb{R}^N)))$, for all admissible pairs (q,r), with norm depending only on q. Let $\varphi \in C([0,T], H^1(\mathbb{R}^N)) \cap C([0,T], L^{r'}(\mathbb{R}^N))$ be such that $\|\varphi\|_{L^{q'}((0,T),L^{r'})} \le 1$. In particular, $\Phi_f \in C([0,T], L^2(\mathbb{R}^N))$, and (by Step 4) $\|\Psi_\varphi\|_{L^\infty((0,T), L^2)} \le C(q)$. We have

$$\int_0^T \operatorname{Re} \int_{\mathbb{R}^N} \Phi_f \overline{\varphi}\,dx\,dt = \int_0^T \langle \Phi_f(t), \varphi(t) \rangle\,dt$$

$$= \int_0^T \int_0^t \langle \mathcal{T}(t-s)f(s), \varphi(t) \rangle\,ds\,dt$$

$$= \int_0^T \int_0^t \langle f(s), \mathcal{T}(s-t)\varphi(t) \rangle\,ds\,dt$$

$$= \int_0^T \int_s^T \langle f(s), \mathcal{T}(s-t)\varphi(t) \rangle\,dt\,ds$$

$$= \int_0^T \langle f(s), \Psi_\varphi(s) \rangle\,ds.$$

Applying the Cauchy–Schwarz inequality, and then Step 4, it follows that

$$\left| \int_0^T \mathrm{Re} \int_{\mathbb{R}^N} \Phi_f \overline{\varphi} \, dx \, dt \right| \leq \|f\|_{L^1((0,T),L^2)} \|\Psi_\varphi\|_{L^\infty((0,T),L^2)}$$
$$\leq C(q) \|f\|_{L^1((0,T),L^2)} \|\varphi\|_{L^{q'}((0,T),L^{r'})}$$
$$\leq C(q) \|f\|_{L^1((0,T),L^2)}.$$

Then it suffices to apply Remark 7.3.3.

Step 6. Conclusion. Let (γ, ρ) be an admissible pair. By Steps 1 and 3, Φ is a continuous operator $L^{\gamma'}((0,T), L^{\rho'}(\mathbb{R}^N)) \to C([0,T], L^2(\mathbb{R}^N))$ and $L^{\gamma'}((0,T), L^{\rho'}(\mathbb{R}^N)) \to L^{\gamma}((0,T), L^{\rho}(\mathbb{R}^N))$. Let (q, r) be an admissible pair such that $2 \leq r \leq \rho$, and let $\theta \in [0,1]$ be such that

$$\frac{1}{r} = \frac{\theta}{\rho} + \frac{1-\theta}{2} \quad \text{and} \quad \frac{1}{q} = \frac{\theta}{\gamma} + \frac{1-\theta}{\infty}.$$

Applying Hölder's inequality in space and then in time, we obtain

$$\|\Phi_f\|_{L^q((0,T),L^r)} \leq \|\Phi_f\|_{L^\gamma((0,T),L^\rho)}^\theta \|\Phi_f\|_{L^\infty((0,T),L^2)}^{1-\theta} \leq C \|f\|_{L^{\gamma'}((0,T),L^{\rho'})},$$

where C depends only on γ and q. Then, Φ is continuous $L^{\gamma'}((0,T), L^{\rho'}(\mathbb{R}^N)) \to L^q((0,T), L^r(\mathbb{R}^N))$. Now let (q, r) be an admissible pair such that $\rho < r$, and let $\mu \in (0, 1)$ be such that

$$\frac{1}{\gamma'} = \frac{\mu}{1} + \frac{1-\mu}{q'} \quad \text{and} \quad \frac{1}{\rho'} = \frac{\mu}{2} + \frac{1-\mu}{r'}.$$

By Steps 1 and 5, Φ is a continuous operator $L^{q'}((0,T), L^{r'}(\mathbb{R}^N)) \to L^q((0,T), L^r(\mathbb{R}^N))$ and $L^1((0,T), L^2(\mathbb{R}^N)) \to L^q((0,T), L^r(\mathbb{R}^N))$. By interpolation (see Bergh and Löfström [1], Thm. 5.1.2 p. 107), Φ is a continuous operator $L^\sigma((0,T), L^\delta(\mathbb{R}^N)) \to L^q((0,T), L^r(\mathbb{R}^N))$, for all (σ, δ) such that

$$\frac{1}{\sigma} = \frac{\theta}{1} + \frac{1-\theta}{q'} \quad \text{and} \quad \frac{1}{\delta} = \frac{\theta}{2} + \frac{1-\theta}{r'},$$

with $\theta \in (0, 1)$. Choosing $\theta = \mu$, we deduce that $\Phi \in \mathcal{L}(L^{\gamma'}((0,T), L^{\rho'}(\mathbb{R}^N)), L^q((0,T), L^r(\mathbb{R}^N)))$, which completes the proof. $\square$

Corollary 7.3.5. Let (γ, ρ) be an admissible pair. Let $f \in C([0,T], L^2(\mathbb{R}^N)) \cap W^{1,\gamma'}((0,T), L^{\rho'}(\mathbb{R}^N))$. Then $\mathcal{F}_f \in C([0,T], H^2(\mathbb{R}^N)) \cap C^1([0,T], L^2(\mathbb{R}^N))$.

Proof. We have in particular $f \in W^{1,1}([0,T], H^{-1}(\mathbb{R}^N))$, and so (applying Proposition 4.1.6) $\mathcal{F}_f \in C([0,T], H^1(\mathbb{R}^N)) \cap C^1([0,T], H^{-1}(\mathbb{R}^N))$ and

$$\frac{d}{dt} \Phi_f = i \Delta \Phi_f + f, \tag{7.21}$$

for all $t \in [0,T]$. In addition, we have

$$\Phi_f(t) = \int_0^t \mathcal{T}(s)f(t-s)\,ds,$$

and so

$$\frac{d}{dt}\Phi_f(t) = \mathcal{T}(t)f(0) + \int_0^t \mathcal{T}(s)f'(t-s)\,ds = \Phi_{f'}(t).$$

This relation is clear if $f \in C^1([0,T], L^{r'}(\mathbb{R}^N)) \subset C^1([0,T], H^{-1}(\mathbb{R}^N))$, and is obtained by density in the general case. Applying Proposition 7.3.4, it follows that $\Phi_f \in C^1([0,T], L^2(\mathbb{R}^N))$. By (7.21), we have $\Delta\Phi_f \in C([0,T], L^2(\mathbb{R}^N))$; and so $\Phi_f \in C([0,T], H^2(\mathbb{R}^N))$. □

In the spirit of Proposition 7.3.4, we give the following result.

Proposition 7.3.6. *Let $\varphi \in L^2(\mathbb{R}^N)$. For all admissible pairs (q,r) we have $\mathcal{T}(\cdot)\varphi \in L^q(\mathbb{R}, L^r(\mathbb{R}^N))$. Moreover, there exists a constant C, depending only on q, such that*

$$\|\mathcal{T}(\cdot)\varphi\|_{L^q(\mathbb{R},L^r)} \leq C(q)\|\varphi\|_{L^2},$$

for all $\varphi \in L^2(\mathbb{R}^N)$.

Proof. The proof is similar to that of Proposition 7.3.4, and so we only sketch it briefly. Set

$$\Lambda_f(t) = \int_{-\infty}^{+\infty} \mathcal{T}(t-s)f(s)\,ds, \quad \Gamma_f = \int_{-\infty}^{+\infty} \mathcal{T}(-t)f(t)\,dt.$$

We show that (see Step 1 of the proof of Proposition 7.3.4)

$$\|\Lambda_f\|_{L^q(\mathbb{R},L^r)} \leq C(q)\|f\|_{L^{q'}(\mathbb{R},L^{r'})}.$$

Next we deduce (see Step 3 of the proof of Proposition 7.3.4) that

$$\|\Gamma_f\|_{L^2} \leq \|f\|_{L^{q'}(\mathbb{R},L^{r'})}.$$

We conclude (as in Step 5 of the proof of Proposition 7.3.4) by noting that

$$\left|\int_{-\infty}^{+\infty}\langle\mathcal{T}(t)\varphi,\psi(t)\rangle\right| = \left|\langle\varphi,\int_{-\infty}^{+\infty}\mathcal{T}(-t)\psi(t)\rangle\right| \leq C(q)\|\varphi\|_{L^2}\|\psi\|_{L^{q'}(\mathbb{R},L^{r'})},$$

and applying Remark 7.3.3. □

7.4. The non-linear Schrödinger equation in $\mathbb{R}^N$: local existence

We extend the result of Theorem 7.2.1 to more general non-linearities. From now on, g is a function $\mathbb{R}_+ \to \mathbb{R}$ such that $g(0) = 0$ and such that there exist $K < \infty$ and $\alpha \geq 0$ with $(N-2)\alpha \leq 4$, such that

$$|g(y) - g(x)| \leq K(1 + |x|^\alpha + |y|^\alpha)|y - x|,$$

for all $x, y \in \mathbb{R}_+$. We extend g to the complex plane by formula (7.8) and we define the function $G : \mathbb{C} \to \mathbb{R}_+$ by formula (7.9). g defines a mapping $L_{\text{loc}}^{\alpha+1} \to L_{\text{loc}}^1$ by formula (7.10). We verify (see §6.1.2) that g is continuous from $H^1(\mathbb{R}^N)$ to $H^{-1}(\mathbb{R}^N)$, and that V defined by formula (7.11) is continuous on $H^1(\mathbb{R}^N)$. Finally, we define the functional E on $H^1(\mathbb{R}^N)$ by formula (7.12). The main result of this section is the following theorem.

Theorem 7.4.1. *There exists a function* $T : H^1(\mathbb{R}^N) \to (0, \infty]$ *that satisfies the following properties. For all* $\varphi \in H^1(\mathbb{R}^N)$, *there exists a function* $u \in C([0, T(\varphi)), H^1(\mathbb{R}^N))$ *which is the unique solution of (7.1)–(7.3) for all* $T < T(\varphi)$. *In addition,*

(i) if $T(\varphi) < \infty$, *then* $\|u(t)\|_{H^1} \to \infty$;

(ii) $\|u(t)\|_{L^2} = \|\varphi\|_{L^2}$, *for all* $t \in [0, T(\varphi))$;

(iii) $E(u(t)) = E(\varphi)$, *for all* $t \in [0, T(\varphi))$;

(iv) if $\varphi \in H^2(\mathbb{R}^N)$, *then* $u \in C([0, T(\varphi)), H^2(\mathbb{R}^N)) \cap C^1([0, T(\varphi)), L^2(\mathbb{R}^N))$.

Finally, the function $T : H^1(\mathbb{R}^N) \to (0, \infty]$ *is lower semicontinuous, and* u *depends continuously on* φ *in the following sense: if* $\varphi_n \to \varphi$ *in* $H^1(\mathbb{R}^N)$, *and if* u_n *and* u *denote the corresponding solutions of (7.1)–(7.3), then* $u_n \to u$ *in* $C([0, T], H^1(\mathbb{R}^N))$, *for all* $T < T(\varphi)$.

Remark. Property (iv) above is a regularity result. As well as being interesting in its own right, this result will be indirectly useful in §7.6 (blow-up in finite time) and §7.7 (behaviour at infinity) to establish the fundamental identity (7.58). This property has been included in Theorem 7.4.1 since its proof requires the approximation argument which we use for the local existence in $H^1(\mathbb{R}^N)$. Doing this, the proof of Theorem 7.4.1 turns out to be much more complicated. Consequently, at the first reading, the reader should omit Lemmas 7.4.7, 7.4.9, and 7.4.10, Corollary 7.4.8, Steps 3 and 5 of the proof of Proposition 7.4.12, and Step 3 of the proof of Corollary 7.4.13 (identified by an asterisk). These results, whose aim is to prove property (iv), are somewhat technical and may obscure the proof of the local existence in $H^1(\mathbb{R}^N)$.

In fact, the proof of the local existence itself is not easy. We sketch it briefly and point out the difficulties.

In §7.2, it appears clearly that the methods developed so far do not allow to solve (7.1)–(7.3) if g is superlinear (see Remark 7.2.3). In $\mathbb{R}^N$, we can get round this difficulty by using the inequalities proved in §7.3. Recall, however, that these inequalities are specific to $\mathbb{R}^N$.

The idea of the proof of Theorem 7.4.1 is the following. We approximate g in a convenient sense by a sequence $(g_m)_{m\geq 0}$ of globally Lipschitz non-linearities, which satisfy the same assumptions as g uniformly with respect to m. In a preliminary section (§7.4.1), we apply the estimates of §7.3 to establish various inequalities for g_m and for g. In general, they derive from Proposition 7.3.4 and Hölder's inequality. We apply these inequalities in §7.4.2. First, (Lemma 7.4.11), we obtain immediately a uniqueness result (note that, at this stage, we do not use approximation but only Proposition 7.3.4 and Hölder's inequality). Next (Proposition 7.4.12) applying Theorem 7.2.1 to problem (7.1)–(7.3) with g replaced by g_m, we build solutions u_m of the approximate problems. On a small time interval, we estimate these solutions u_m, uniformly with respect to m, using mainly conservation of energy. These estimates allow us to pass to the limit as $m \to \infty$, and to obtain a local solution u. Solving the backward problem, we demonstrate that energy is conserved (Corollary 7.4.13). Finally, we complete the proof showing alternative (i) and continuous dependence. To establish property (iv), we may suppose that $\varphi \in H^2(\mathbb{R}^N)$ and deduce estimates of u_m in $H^2(\mathbb{R}^N)$ which are independent of m. We then pass to the limit in these estimates as $m \to \infty$.

7.4.1. Some estimates

The following notation will be useful in what follows. We define $k \in C(\mathbb{C}, \mathbb{C})$ by

$$k(z) = \begin{cases} g(z), & \text{if } |z| \leq 1; \\ zg(1), & \text{if } |z| \geq 1. \end{cases}$$

for all $m \in \mathbb{N}$, we define $g_m \in C(\mathbb{C}, \mathbb{C})$ and $h_m \in C(\mathbb{C}, \mathbb{C})$ by

$$g_m(z) = \begin{cases} g(z), & \text{if } |z| \leq m; \\ \dfrac{z}{m}g(m), & \text{if } |z| \geq m; \end{cases}$$

$$h_m(z) = g_m(z) - k(z).$$

(We have in particular $g_1 = k$.) Also, let

$$g_\infty = g;$$
$$h_\infty = g - k.$$

For $m \in \mathbb{N} \cup \{\infty\}$, we define $G_m \in C(\mathbb{C}, \mathbb{R})$ by

$$G_m(z) = \int_0^{|z|} g_m(s)\,ds.$$

Observe that for all $m \in \mathbb{N}$, g_m is globally Lipschitz continuous. For all $u \in H^1(\mathbb{R}^N)$, we set

$$V_m(u) = -\int G_m(|u|);$$

$$E_m(u) = \frac{1}{2}\int_{\mathbb{R}^N} |\nabla u|^2\, dx + V_m(u).$$

We set $V_\infty = V$ and $E_\infty = E$. Finally, for all $T > 0$, $u \in L^\infty((0,T), H^1(\mathbb{R}^N))$ and $m \in \mathbb{N} \cup \{\infty\}$, we set

$$\mathcal{G}_m(u)(t) = \int_0^t \mathcal{T}(t-s)g_m(u(s))\, ds;$$

$$\mathcal{H}_m(u)(t) = \int_0^t \mathcal{T}(t-s)h_m(u(s))\, ds;$$

$$\mathcal{K}(u)(t) = \int_0^t \mathcal{T}(t-s)k(u(s))\, ds;$$

for all $t \in [0,T]$, and we let $\mathcal{H}_\infty = \mathcal{H}$, $\mathcal{G}_\infty = \mathcal{G}$.

By possibly modifying the value of K, we readily verify the following inequalities:

$$|g_m(z_2) - g_m(z_1)| \le K(1 + |z_2|^\alpha + |z_1|^\alpha)|z_2 - z_1|; \tag{7.22}$$

$$|g_m(z) - g(z)| \le K|z|^{\alpha+1}; \tag{7.23}$$

$$|h_m(z_2) - h_m(z_1)| \le K(|z_2|^\alpha + |z_1|^\alpha)|z_2 - z_1|; \tag{7.24}$$

$$|k(z_2) - k(z_1)| \le K|z_2 - z_1|; \tag{7.25}$$

for all $m \in \mathbb{N} \cup \{\infty\}$ and all $z, z_1, z_2 \in \mathbb{C}$. In what follows we set $\tau = \alpha + 2$ and $\sigma = 4(\alpha+2)/(N\alpha)$, so that (σ, τ) is admissible.

Lemma 7.4.2. *Let $M < \infty$. There exists $C(M) < \infty$ and $\nu > 0$ such that*

$$\|k(v) - k(u)\|_{L^2} \le C(M)\|v - u\|_{L^2}; \tag{7.26}$$

$$\|h_m(v) - h_m(u)\|_{L^{\tau'}} \le C(M)\|v - u\|_{L^\tau}; \tag{7.27}$$

$$\|g_m(u) - g(u)\|_{L^{\tau'}} \le C(M)m^{-\nu}; \tag{7.28}$$

for all $u, v \in H^1(\mathbb{R}^N)$ such that $\|u\|_{H^1} \le M$ and $\|v\|_{H^1} \le M$ and for all $m \in \mathbb{N}$.

Proof. (7.26) is an immediate consequence of (7.25). (7.27) is a consequence of (7.24), Hölder's inequality, and of the embedding $H^1(\mathbb{R}^N) \hookrightarrow L^\tau(\mathbb{R}^N)$. Finally, we observe that

$$|g_m(z) - g(z)| \le \begin{cases} 0, & \text{if } |z| \le m; \\ K|z|^{\alpha+1}, & \text{if } |z| \ge m. \end{cases}$$

Let $r > \tau$ be such that $(N-2)r < 2N$. Then we have

$$\int_{\mathbb{R}^N} |g_m(u) - g(u)|^{\tau'} \le K \int_{|u|\ge m} |u|^{\tau} \le Km^{-(r-\tau)} \int_{|u|\ge m} |u|^r$$
$$\le CKm^{-(r-\tau)} \|u\|_{H^1}^r.$$

From this, we deduce (7.28). □

Lemma 7.4.3. Let $M < \infty$. There exist $C(M) < \infty$ and $\mu > 0$ such that

$$|V_m(v) - V_m(u)| \le C(M) \left(\|v-u\|_{L^2} + \|v-u\|_{L^2}^{1-2/\sigma} \right); \quad (7.29)$$
$$|V_m(u) - V(u)| \le C(M) m^{-\mu}; \quad (7.30)$$

for all $u, v \in H^1(\mathbb{R}^N)$ such that $\|u\|_{H^1} \le M$ and $\|v\|_{H^1} \le M$ and all $m \in \mathbb{N} \cup \{\infty\}$.

Proof. We observe that, by (7.23),

$$|G_m(z_2) - G_m(z_1)| \le K(|z_2| + |z_1| + |z_2|^{\alpha+1} + |z_1|^{\alpha+1})|z_2 - z_1|;$$

and so, by Hölder's inequality,

$$|V_m(v) - V_m(u)| \le C(\|u\|_{L^2} + \|v\|_{L^2})\|v-u\|_{L^2} + C(\|u\|_{L^r}^{\alpha+1} + \|v\|_{L^r}^{\alpha+1})\|v-u\|_{L^r}.$$

On the other hand, for all $u \in H^1(\mathbb{R}^N)$, we have (Theorem 1.3.4)

$$\|u\|_{L^r} \le \|u\|_{H^1}^a \|u\|_{L^2}^{1-a}, \quad \text{with } a = N\left(\frac{1}{2} - \frac{1}{r}\right) = \frac{2}{\sigma}; \quad (7.31)$$

and hence (7.29). (7.30) is proved as (7.28), observing that

$$|G_m(z) - G(z)| \le \begin{cases} 0, & \text{if } |z| \le m; \\ K|z|^r, & \text{if } |z| \ge m; \end{cases}$$

for all $m \in \mathbb{N}$. □

Lemma 7.4.4. Let $M < \infty$ and let (q, r) be an admissible pair. There exist $C(M, q) < \infty$ and $\nu > 0$ such that

$$\|\mathcal{K}(v) - \mathcal{K}(u)\|_{L^q((0,T),L^r)} \le C(M,q)T\|v-u\|_{L^\infty((0,T),L^2)}; \quad (7.32)$$
$$\|\mathcal{H}_m(v) - \mathcal{H}_m(u)\|_{L^q((0,T),L^r)} \le C(M,q)T^{1-2/\sigma}\|v-u\|_{L^\sigma((0,T),L^r)}; \quad (7.33)$$
$$\|\mathcal{G}_m(u) - \mathcal{G}(u)\|_{L^q((0,T),L^r)} \le C(M,q)T^{\frac{1}{q}}m^{-\nu}; \quad (7.34)$$

for all $T > 0$, all $m \in \mathbb{N} \cup \{\infty\}$ and all $u, v \in L^\infty((0,T), H^1(\mathbb{R}^N))$ such that $\|u\|_{L^\infty((0,T), H^1)} \le M$ and $\|v\|_{L^\infty((0,T), H^1)} \le M$.

Proof. We apply (7.20), next Hölder's inequality on $(0, T)$, and finally inequalities (7.26), (7.27), or (7.28). □

Lemma 7.4.5. Let $T > 0$ and let $u \in L^\infty((0,T), H^1(\mathbb{R}^N)) \cap W^{1,\infty}((0,T), H^{-1}(\mathbb{R}^N))$. Set

$$K = \max\{\|u\|_{L^\infty((0,T), H^1)}, \|u'\|_{L^\infty((0,T), H^{-1})}\}.$$

Then $u \in C([0, T], L^2(\mathbb{R}^N))$ and

$$\|u(t) - u(s)\|_{L^2} \le 2K|t - s|^{\frac{1}{2}},$$

for all $t, s \in [0, T]$.

Proof. We have $u \in C([0, T], H^{-1}(\mathbb{R}^N))$ (Corollary 1.4.36), and so $u : [0, T] \to H^1(\mathbb{R}^N)$ is weakly continuous. In addition, we have, for all $t, s \in [0, T]$,

$$\|u(t) - u(s)\|_{L^2}^2 = \langle u(t) - u(s), u(t) - u(s)\rangle_{H^{-1}, H^1} \le 2K\|u(t) - u(s)\|_{H^{-1}}$$

$$\le 2K \int_s^t \|u'(\sigma)\|_{H^{-1}} \, d\sigma \le 2K^2|t - s|;$$

hence the result. □

Lemma 7.4.6. Let $T > 0$, $u \in L^\infty((0,T), H^1(\mathbb{R}^N)) \cap C([0, T], L^2(\mathbb{R}^N))$. Let $r \ge 2$ be such that $(N - 2)r < 2N$. Then $u \in C([0, T], L^r(\mathbb{R}^N))$ and

$$\|u\|_{L^\infty((0,T), L^r)} \le \|u\|_{L^\infty((0,T), H^1)}^{N(1/2 - 1/r)} \|u\|_{L^\infty((0,T), L^2)}^{1 - N(1/2 - 1/r)}, \tag{7.35}$$

where C does not depend on u.

Proof. Applying (7.31) to $w = u(t + h) - u(t)$, we obtain the continuity and then, letting $w = u(t)$, we obtain (7.35). □

Similarly, we show the following lemma.

Lemma 7.4.7 (*) Let $T > 0$, $u \in L^\infty((0,T), H^2(\mathbb{R}^N)) \cap C([0, T], L^2(\mathbb{R}^N))$. Let $r \ge 2$ be such that $(N - 4)r < 2N$. Then $u \in C([0, T], L^r(\mathbb{R}^N))$.

Corollary 7.4.8 (*). If $u \in L^\infty((0,T), H^2(\mathbb{R}^N)) \cap C([0, T], L^2(\mathbb{R}^N))$, then $g(u) \in C([0, T], L^2(\mathbb{R}^N))$.

Proof. We have in particular $u \in C([0,T], L^{2(\alpha+1)}(\mathbb{R}^N)) \cap C([0,T], L^2(\mathbb{R}^N))$, and the result follows readily. □

Lemma 7.4.9 (*). *For all M, there exists $C(M)$ such that*

$$\|g_m(u)\|_{L^2} \leq \frac{1}{2}\|\triangle u\|_{L^2} + C(M),$$

for all $m \in \mathbb{N}$ and all $u \in H^1(\mathbb{R}^N)$ such that $\|u\|_{H^1} \leq M$.

Proof. It suffices to estimate h_m. If $N \leq 2$, we have $H^1(\mathbb{R}^N) \hookrightarrow L^{2(\alpha+1)}(\mathbb{R}^N)$, and then $\|h_m(u)\|_{L^2} \leq C(M)$. If $N \geq 3$, we may suppose that $\alpha \geq 2/(N-2)$, so that $2(\alpha+1) \geq 2N/(N-2)$. In that case, we have (Theorem 1.3.4)

$$\begin{aligned}
\|h_m(u)\|_{L^2} &\leq C\|u\|_{L^{2(\alpha+1)}}^{\alpha+1} \\
&\leq C\|\triangle u\|_{L^2}^{a(\alpha+1)}\|u\|_{L^{\frac{2N}{N-2}}}^{(1-a)(\alpha+1)} \\
&\leq C\|\triangle u\|_{L^2}^{a(\alpha+1)}M^{(1-a)(\alpha+1)},
\end{aligned}$$

with $a/N = 1/2 - 1/N - 1/(2(\alpha+1))$. In particular, we have $a(\alpha+1) < 1$; hence the result. □

Lemma 7.4.10 (*) *For all M, there exists $C(M)$ with the following properties. For all $T > 0$, all $m \in \mathbb{N} \cup \{\infty\}$ and all $u \in L^\infty((0,T), H^1(\mathbb{R}^N)) \cap W^{1,\infty}((0,T), L^2(\mathbb{R}^N)) \cap W^{1,\sigma}((0,T), L^r(\mathbb{R}^N))$ such that $\|u\|_{L^\infty((0,T),H^1)} \leq M$, we have $k(u) \in W^{1,\infty}((0,T), L^2(\mathbb{R}^N))$ and $h_m(u) \in W^{1,\sigma'}((0,T), L^{r'}(\mathbb{R}^N))$; in addition,*

$$\|k(u)'\|_{L^1((0,T),L^2)} \leq C(M)T\|u'\|_{L^\infty((0,T),L^2)}; \tag{7.36}$$
$$\|h_m(u)'\|_{L^{\sigma'}((0,T),L^{r'})} \leq C(M)T^{1-2/\sigma}\|u'\|_{L^\sigma((0,T),L^r)}. \tag{7.37}$$

Proof. By Theorem 1.4.40, u is Lipschitz $[0,T] \to L^2(\mathbb{R}^N)$; and then so is $k(u)$, by (7.26). Applying Theorem 1.4.40, we thus obtain $k(u)' \in L^\infty((0,T), L^2(\mathbb{R}^N))$ and

$$\|k(u)'\|_{L^\infty((0,T),L^2)} \leq C(M)\|u'\|_{L^\infty((0,T),L^2)};$$

hence (7.36). Applying (7.27) we show that $h_m(u) \in W^{1,\sigma}((0,T), L^{r'}(\mathbb{R}^N))$ and that

$$\|h_m(u)'\|_{L^\sigma((0,T),L^{r'})} \leq C(M)\|u'\|_{L^\sigma((0,T),L^r)};$$

(7.37) follows. □

7.4.2. Proof of Theorem 7.4.1

We begin with a uniqueness result.

Lemma 7.4.11. *Let $T > 0$ and $\varphi \in H^1(\mathbb{R}^N)$. Let $u, v \in L^\infty((0,T), H^1(\mathbb{R}^N))$ be two solutions of (7.4). Then $u = v$.*

Proof. Observe first that by Lemmas 7.1.2 and 7.4.5, $u, v \in C([0,T], L^2(\mathbb{R}^N))$. Let $\theta = \sup\{t \in [0,T]; u = v \text{ on } [0,t]\}$. If $\theta = T$, we have $u = v$ on $[0,T]$. If $\theta < T$, we may suppose $\theta = 0$, using the transformation $t \to t - \theta$. Set $M = \max\{\|u\|_{L^\infty((0,T),H^1)}, \|v\|_{L^\infty((0,T),H^1)}\}$, and let (q, r) be an admissible pair. We have

$$u - v = \mathrm{i}(\mathcal{K}u - \mathcal{K}v) + \mathrm{i}(\mathcal{H}u - \mathcal{H}v),$$

and applying (7.32)–(7.33), we obtain, for all $t \in (0, T]$

$$\|u-v\|_{L^q((0,t),L^r)} \leq C(M,q)(t\|u-v\|_{L^\infty((0,t),L^2)} + t^{1-2/\sigma}\|u-v\|_{L^\sigma((0,t),L^\tau)}).$$

Choosing successively $(q, r) = (\infty, 2)$ and $(q, r) = (\sigma, \tau)$, making the sum and then taking t sufficiently small, we obtain

$$\|u-v\|_{L^\infty((0,t),L^2)} + \|u-v\|_{L^\sigma((0,t),L^\tau)} = 0;$$

We therefore have the desired contradiction with the hypothesis $\theta = 0$. $\square$

Proposition 7.4.12. *For all M, there exists $T_M > 0$ with the following properties. For all $\varphi \in H^1(\mathbb{R}^N)$ such that $\|\varphi\|_{H^1} \leq M$, there exists a solution $u \in L^\infty((0,T_M); H^1(\mathbb{R}^N)) \cap W^{1,\infty}((0,T_M); H^{-1}(\mathbb{R}^N))$ of (7.4). This solution satisfies:*

(i) $\|u\|_{L^\infty((0,T_M);H^1)} \leq 2M$;

(ii) $\|u(t)\|_{L^2} = \|\varphi\|_{L^2}$, *for all* $t \in [0, T_M]$;

(iii) $E(u(t)) \leq E(\varphi)$, *for all* $t \in [0, T_M]$;

(iv) *if* $\varphi \in H^2(\mathbb{R}^N)$, *then* $u \in C([0, T_M], H^2(\mathbb{R}^N)) \cap C^1([0, T_M], L^2(\mathbb{R}^N))$.

In addition, if u and v are the solutions corresponding to the initial data φ and ψ, we have:

(v) $\|u - v\|_{L^\infty((0,T_M);L^2)} \leq K\|\varphi - \psi\|_{L^2}$,

where K is a constant which does not depend on M.

Proof. We proceed in six steps. We consider $M > 0$ and $\varphi \in H^1(\mathbb{R}^N)$ such that $\|\varphi\|_{H^1} \leq M$.

Step 1. Construction of approximate solutions. For all $m \in \mathbb{N}$, the function g_m is globally Lipschitz continuous. Thus we may apply Theorem 7.2.1. There exists a unique function $u_m \in C([0,\infty), H^1(\mathbb{R}^N)) \cap C^1([0,\infty), H^{-1}(\mathbb{R}^N))$, a solution of the following equation

$$u_m(t) = \mathcal{T}(t)\varphi + i\mathcal{G}_m(u_m)(t), \tag{7.38}$$

for all $t \geq 0$. In addition, we have

$$\|u_m(t)\|_{L^2} = \|\varphi\|_{L^2}; \tag{7.39}$$
$$E_m(u_m(t)) = E_m(\varphi); \tag{7.40}$$

for all $t \geq 0$.

Step 2. A *priori* estimates on the solutions. Let

$$\tau_m = \sup\{t \geq 0;\ \|u_m\|_{C([0,t], H^1)} \leq 2M\}.$$

Our intention is to show that there exists T_M, depending only on M, such that $\tau_m \geq T_M$. To see this, assume that $\tau_m < \infty$. In that case, we have

$$\|u_m(\tau_m)\|_{H^1} = 2M. \tag{7.41}$$

But, by (7.39) and (7.40), we have

$$\|u_m(t)\|_{H^1}^2 = \|\varphi\|_{H^1}^2 + 2\left(V_m(u_m(t)) - V_m(\varphi)\right). \tag{7.42}$$

On the other hand, applying Lemma 7.1.1, (7.26), (7.27), and Sobolev's embeddings, we obtain, for all $t \in [0, \tau_m]$,

$$\|u_m'\|_{L^\infty((0,t), H^{-1})} \leq \|\Delta u_m\|_{L^\infty((0,t), H^{-1})} + \|k(u_m)\|_{L^\infty((0,t), H^{-1})}$$
$$+ \|h_m(u_m)\|_{L^\infty((0,t), H^{-1})}$$
$$\leq \|u_m\|_{L^\infty((0,t), H^1)} + \|k(u_m)\|_{L^\infty((0,t), L^2)}$$
$$+ C'\|h_m(u_m)\|_{L^\infty((0,t), L^{r'})}$$
$$\leq 2M + 2MC(M) + 2MC''C(M).$$

By Lemma 7.4.5 and (7.29), it follows that there exists $D(M)$, depending only on M, such that

$$2|V_m(u_m(t)) - V_m(\varphi)| \leq D(M)(t^{1/2} + t^{1/2 - 1/\sigma}),$$

for all $t \in [0, \tau_m]$. Putting this into (7.42) and applying (7.41), we obtain

$$4M^2 \leq M^2 + D(M)(\tau_m^{1/2} + \tau_m^{1/2 - 1/\sigma}).$$

It follows readily that there exists $T_M > 0$ depending only on M, such that $\tau_m \geq T_M$, and so
$$\sup\{\|u_m(t)\|_{H^1}; t \in [0, T_M]\} \leq 2M, \tag{7.43}$$
which is the desired estimate.

Step 3 (*). The case $\varphi \in H^2(\mathbb{R}^N)$. In that case, we know (Theorem 7.2.1) that $u_m \in C([0,\infty), H^2(\mathbb{R}^N)) \cap C^1([0,\infty), L^2(\mathbb{R}^N))$. Since g_m is Lipschitz, we have in particular $g_m(u_m) \in W^{1,\infty}((0, T_M), L^2(\mathbb{R}^N))$ and, for all $t \in [0, T_M]$:
$$u'_m(t) = \mathcal{T}(t)(\mathrm{i}\Delta\varphi + \mathrm{i}g_m(\varphi)) + \mathrm{i}\int_0^t \mathcal{T}(t-s)g_m(u_m)'(s)\,\mathrm{d}s. \tag{7.44}$$

From Propositions 7.3.4 and 7.3.6, it follows that $u_m \in W^{1,\sigma}((0, T_M); L^r(\mathbb{R}^N))$. Then, we write (7.44) as
$$\begin{aligned}u'_m(t) = \mathcal{T}(t)(\mathrm{i}\Delta\varphi + \mathrm{i}g_m(\varphi)) &+ \mathrm{i}\int_0^t \mathcal{T}(t-s)k(u_m)'(s)\,\mathrm{d}s \\ &+ \mathrm{i}\int_0^t \mathcal{T}(t-s)h_m(u_m)'(s)\,\mathrm{d}s.\end{aligned} \tag{7.45}$$

By Lemma 7.4.9, $\Delta\varphi + g_m(\varphi)$ is bounded in $L^2(\mathbb{R}^N)$. We estimate the first term on the right-hand side of (7.45) by Proposition 7.3.6 and the integrals as a consequence of Proposition 7.3.4 and Lemma 7.4.10. It follows that, for all $T \leq T_M$,
$$\begin{aligned}\|u'_m\|_{L^\infty((0,T),L^2)} &+ \|u'_m\|_{L^\sigma((0,T),L^r)} \\ &\leq C(\varphi) + K(M)(T + T^{1-2/\sigma})(\|u'_m\|_{L^\infty((0,T),L^2)} + \|u'_m\|_{L^\sigma((0,T),L^r)}),\end{aligned}$$
where $K(M)$ depends only on M and C depends on $\|\varphi\|_{H^2}$. Choosing if necessary T_M smaller (but depending only on M), we may assume that $K(M)(T_M + T_M^{1-2/\sigma}) \leq 1/2$. We then obtain
$$\|u'_m\|_{L^\infty((0,T_M);L^2)} + \|u'_m\|_{L^\sigma((0,T_M);L^r)} \leq 2C(\varphi). \tag{7.46}$$

According to the equation satisfied by u_m, we also obtain
$$\|\Delta u_m\|_{L^\infty((0,T_M);L^2)} \leq \|u'_m\|_{L^\infty((0,T_M);L^2)} + \|g_m(u_m)\|_{L^\infty((0,T_M);L^2)}.$$

Next, we apply (7.46) and Lemma 7.4.9. It follows that
$$\|\Delta u_m\|_{L^\infty((0,T_M);L^2)} \leq D(\varphi), \tag{7.47}$$
where D depends only on $\|\varphi\|_{H^2}$.

Step 4. Convergence of the sequence $(u_m)_{m\geq 0}$. Let m and p be two integers. We write that, for all $t \geq 0$,

$$u_m(t) - u_p(t) = \mathrm{i}(\mathcal{K}(u_m)(t) - \mathcal{K}(u_p)(t)) + \mathrm{i}(\mathcal{H}_m(u_m)(t) - \mathcal{H}_m(u_p)(t))$$
$$+ \mathrm{i}(\mathcal{H}_m(u_p)(t) - \mathcal{H}_p(u_p)(t)).$$

Apply Lemma 7.4.4 successively with $(q,r) = (\infty, 2)$ and $(q,r) = (\sigma, \tau)$. We deduce that there exists $C(M)$, depending only on M, such that

$$\|u_m - u_p\|_{L^\infty((0,T_M);L^2)} + \|u_m - u_p\|_{L^\sigma((0,T_M);L^\tau)} \leq C(M)(1+T_M)(m^{-\nu}+p^{-\nu})$$
$$+ C(M)\left(T_M + T_M^{1-2/\sigma}\right)(\|u_m - u_p\|_{L^\infty((0,T_M);L^2)} + \|u_m - u_p\|_{L^\sigma((0,T_M);L^\tau)}).$$

Taking possibly T_M smaller (but depending only on M as above) we may suppose that
$$C(M)\left(T_M + T_M^{1-2/\sigma}\right) \leq \frac{1}{2}.$$

Therefore $(u_m)_{m\geq 0}$ is a Cauchy sequence in $L^\infty((0,T_M);L^2(\mathbb{R}^N)) \cap L^\sigma((0,T_M);L^\tau(\mathbb{R}^N))$ and there exists $u \in C([0,T_M],L^2(\mathbb{R}^N)) \cap L^\sigma((0,T_M);L^\tau(\mathbb{R}^N))$ such that

$$u_m \longrightarrow u \quad \text{in } C([0,T_M],L^2(\mathbb{R}^N)) \cap L^\sigma((0,T_M);L^\tau(\mathbb{R}^N)) \text{ as } m \to \infty. \quad (7.48)$$

Furthermore, (7.43) and Corollary 1.4.24 imply that $u \in L^\infty((0,T_M);H^1(\mathbb{R}^N))$ and estimate (i) follows. Invoking Lemma 7.4.4, we may immediately pass to the limit in equation (7.38) and conclude that u is a solution of (7.4) on $[0,T_M]$. The conservation law (ii) is a consequence of (7.39) and (7.48). Lemmas 7.4.3 and (7.48) prove that

$$V_m(u_m(t)) \longrightarrow V(u(t)) \quad \text{as } m \to \infty, \quad (7.49)$$

for all $t \in [0,T_M]$. Next let $t \in [0,T_M]$. Since $u_m(t) \rightharpoonup u(t)$ in $H^1(\mathbb{R}^N)$ as $m \to \infty$, we also have

$$\int_{\mathbb{R}^N} |\nabla u(t)|\, dx \leq \liminf_{m \to \infty} \int_{\mathbb{R}^N} |\nabla u_m(t)|\, dx. \quad (7.50)$$

Applying (7.49) and (7.50), we obtain (iii).

Step 5 (*). Returning to the case $\varphi \in H^2(\mathbb{R}^N)$. By (7.47), u_m is bounded in $L^\infty((0,T_M);H^2(\mathbb{R}^N))$, and so (Corollary 1.4.24) $u \in L^\infty((0,T_M);H^2(\mathbb{R}^N))$. In particular (Corollary 7.4.8), $g(u) \in C([0,T_M],L^2(\mathbb{R}^N))$.

Applying estimate (7.46) and Corollary 1.4.42, we obtain in addition that $u \in W^{1,\infty}((0,T_M);L^2(\mathbb{R}^N)) \cap W^{1,\sigma}((0,T_M);L^\tau(\mathbb{R}^N))$. Applying Lemma 7.4.10, we have $k(u) \in W^{1,\infty}((0,T_M); L^2(\mathbb{R}^N))$ and $h(u) \in W^{1,\sigma'}((0,T_M); L^{\tau'}(\mathbb{R}^N))$. Finally, Corollary 7.3.5 implies $u \in C^1([0,T_M], L^2(\mathbb{R}^N)) \cap C([0,T_M], H^2(\mathbb{R}^N))$.

Step 6. The continuous dependence with respect to φ. Let φ and ψ be such that $\|\varphi\|_{H^1} \leq M$ and $\|\psi\|_{H^1} \leq M$. Let u and v be the corresponding solutions of (7.4). We have

$$u - v = \mathcal{T}(\cdot)(\varphi - \psi) + i(\mathcal{K}u - \mathcal{K}v) + i(\mathcal{H}u - \mathcal{H}v),$$

on $[0, T_M]$. We estimate the first term on the right-hand side with Proposition 7.3.6 and the following two terms by Lemma 7.4.4, by taking successively $(q,r) = (\infty, 2)$ and $(q,r) = (\sigma, \tau)$. We deduce that there exist $K < \infty$ and $C(M)$, depending only on M, such that

$$\|u - v\|_{L^\infty((0,T_M); L^2)} + \|u - v\|_{L^\sigma((0,T_M); L^\tau)} \leq K\|\varphi - \psi\|_{L^2}$$
$$+ C(M)\left(T_M + T_M^{1-2/\sigma}\right)\left(\|u - v\|_{L^\infty((0,T_M); L^2)} + \|u - v\|_{L^\sigma((0,T_M); L^\tau)}\right).$$

Taking T_M smaller if necessary (but depending only on M), we may assume that

$$C(M)\left(T_M + T_M^{1-2/\sigma}\right) \leq \frac{1}{2}.$$

We deduce (v). □

Corollary 7.4.13. *Let $T > 0$, $\varphi \in H^1(\mathbb{R}^N)$, and let $u \in L^\infty((0,T), H^1(\mathbb{R}^N))$ be a solution of (7.4). Then $u \in C([0,T], H^1(\mathbb{R}^N)) \cap C^1([0,T], H^{-1}(\mathbb{R}^N))$, and we have:*

(i) $\|u(t)\|_{L^2} = \|\varphi\|_{L^2}$, *for all $t \in [0,T]$;*

(ii) $E(u(t)) = E(\varphi)$, *for all $t \in [0,T]$.*

In addition, if $\varphi \in H^2(\mathbb{R}^N)$, then $u \in C([0,T], H^2(\mathbb{R}^N)) \cap C^1([0,T], L^2(\mathbb{R}^N))$.

Proof. We proceed in three steps.

Step 1. The conservation laws (i) and (ii). Let u be as in the statement of the corollary. Observe that $u : [0,T] \to H^1(\mathbb{R}^N)$ is weakly continuous. Set $M = \|u\|_{L^\infty((0,T), H^1)}$ and

$$\theta = \sup\{t \in [0,T];\ (i)\ \text{and}\ (ii)\ \text{hold on}\ [0,t]\}.$$

We want to show that $\theta = T$. To do this, we argue by contradiction, assuming that $\theta < T$. Then, we may consider only the case $\theta = 0$, since otherwise we use the transform $t \to t - \theta$. In particular, there exists $\delta \in (0, T_M)$, such that

$$|\|u(\delta)\|_{L^2} - \|\varphi\|_{L^2}| + |E(u(\delta)) - E(\varphi)| > 0, \qquad (7.51)$$

where T_M is given by Proposition 7.4.12. Denote by v the solution of (7.4) given by Proposition 7.4.12. Since $\delta \in (0, T_M)$, we have $u = v$ on $[0, \delta]$ (Lemma 7.4.11). In particular, we have $\|u(\delta)\|_{L^2} = \|\varphi\|_{L^2}$ and $E(u(\delta)) \le E(\varphi)$ ((ii) and (iii) of Proposition 7.4.12); and so (7.51) is equivalent to

$$E(u(\delta)) < E(\varphi). \tag{7.52}$$

Set $w = \overline{u(t - \delta)}$ for all $t \in [0, \delta]$. Since u satisfies (7.6), w satisfies

$$iw_t + \Delta w + g(w) = 0,$$

for almost all $t \in (0, \delta)$. Therefore (Lemma 7.1.2) w is solution of (7.4) on $[0, d]$, with φ replaced by $w(0)$. But $\|w(0)\|_{H^1} = \|u(\delta)\|_{H^1} \le M$. It follows that w coincides with the solution given by Proposition 7.4.12 on $[0, d]$. In particular, we have $E(w(\delta)) \le E(w(0))$, and so $E(\varphi) \le E(u(\delta))$, which contradicts (7.52). Consequently, we have (i) and (ii).

Step 2. The continuity in $H^1(\mathbb{R}^N)$. Since $u : [0, T] \to H^1(\mathbb{R}^N)$ is weakly continuous, (i) implies that $u : [0, T] \to L^2(\mathbb{R}^N)$ is continuous. By (7.29), $V(u) : [0, T] \to \mathbb{R}$ is also continuous. We then deduce from (ii) that $\|u\|_{H^1} : [0, T] \to \mathbb{R}$ is continuous. It is now clear that $u \in C([0, T], H^1(\mathbb{R}^N))$. Since $g : H^1(\mathbb{R}^N) \to H^{-1}(\mathbb{R}^N)$ is continuous, we have $u \in C([0, T], H^1(\mathbb{R}^N)) \cap C^1([0, T], H^{-1}(\mathbb{R}^N))$.

Step 3 (*). The $H^2(\mathbb{R}^N)$ regularity. Let $m \in \mathbb{N}$ and $\theta > 0$ be such that $\theta \le T_M$ and $m\theta = T$. u coincides with the solution of (7.4) given by Proposition 7.4.12 on $[0, \theta]$, and so we have $u \in C([0, T], H^2(\mathbb{R}^N)) \cap C^1([0, T], L^2(\mathbb{R}^N))$. Iterate, replacing φ successively by $u(j\theta)$, $1 \le j \le m - 1$, in order to obtain $u \in C([0, T], H^2(\mathbb{R}^N)) \cap C^1([0, T], L^2(\mathbb{R}^N))$. □

End of the proof of Theorem 7.4.1. Using Proposition 7.4.12 and Lemma 7.4.11, and arguing as in the proof of Proposition 4.3.4, we show the existence of a maximal solution which satisfies (i). Properties (ii), (iii), and (iv) are consequences of Corollary 7.4.13. It remains to show the continuous dependence. To see this, we consider $\varphi \in H^1(\mathbb{R}^N)$, and a sequence $(\varphi_m)_{m \ge 0}$, such that $\varphi_m \to \varphi$ in $H^1(\mathbb{R}^N)$, as $m \to \infty$. Let u and u_m be the corresponding maximal solutions of (7.4). It suffices to show that, for all $T \in [0, T(\varphi))$, we have $T(\varphi_m) > T$ for m large enough, and that $u_m \to u$ in $C([0, T], H^1(\mathbb{R}^N))$, as $m \to \infty$. Set

$$\begin{aligned} M &= 4\|u\|_{L^\infty((0,T),H^1)}; \\ \theta(m) &= \sup\{t \in [0, T(\varphi_m)), t \le T;\ \|u_m\|_{L^\infty((0,t),H^1)} \le M\}. \end{aligned}$$

Let $k \in \mathbb{N}$ and $\delta > 0$ be such that $\delta \le T_M$ and $k\delta = T$. Applying Proposition 7.4.12, and since $\theta(m) \le T$, we easily obtain that

$$\|u - u_m\|_{L^\infty(0,\theta(m);L^2)} \le K^k \|\varphi - \varphi_m\|_{L^2}. \tag{7.53}$$

Let us show that $\theta(m) = T$, for m large enough. Without loss of generality, we may assume that $\theta(m) \to \theta \in [T_M, T]$, as $m \to \infty$. Let $t \in (0, \theta)$. $u_m(t)$ is defined for m large enough and, according to (7.53), we have $u_m(t) \to u(t)$ in $L^2(\mathbb{R}^N)$, as $m \to \infty$. On the other hand, we have $E(u_m(t)) = E(\varphi_m) \to E(\varphi) = E(u(t))$. It follows (see Step 2 of the proof of Corollary 7.4.13) that $u_m(t) \to u(t)$ in $H^1(\mathbb{R}^N)$, as $m \to \infty$. In particular, we have $\|u_m(t)\|_{H^1} \leq M/2$ for m large enough, and so (Proposition 7.4.12(i)) $\theta(m) \geq \min\{T, t + T_M/2\}$. $t < \theta$ being arbitrary, it follows that $\theta(m) = T$, for m large enough.

Thus, it remains to show that $u_m \to u$ in $C([0, T], H^1(\mathbb{R}^N))$, as $m \to \infty$. We argue by contradiction, and we suppose that there exists a sequence $(t_m)_{m \geq 0}$ in $[0, T]$ and $\varepsilon > 0$, such that $\|u_m(t_m) - u(t_m)\|_{H^1} \geq \varepsilon$. We may assume $t_m \to t \in [0, T]$, which implies that $\|u_m(t_m) - u(t)\|_{H^1} \geq \varepsilon/2$. By (7.53) and since $u \in C([0, T], L^2(\mathbb{R}^N))$, we have $u_m(t_m) \to u(t)$ in $L^2(\mathbb{R}^N)$, as $m \to \infty$. Furthermore, we have $E(u_m(t_m)) = E(\varphi_m) \to E(\varphi) = E(u(t))$. It follows (see Step 2 of the proof of Corollary 7.4.13) that $u_m(t) \to u(t)$ in $H^1(\mathbb{R}^N)$, as $m \to \infty$; hence the contradiction. This completes the proof of Theorem 7.4.1. □

7.5. The non-linear Schrödinger equation in $\mathbb{R}^N$: global existence

We suppose that $\Omega = \mathbb{R}^N$ and that g satisfies the assumptions of Theorem 7.4.1. As for the heat and Klein–Gordon equations, we are going to establish two kinds of results: global existence for all initial data when g satisfies some additional growth assumptions and global existence for small initial data, without any additional hypothesis on g.

Proposition 7.5.1. *Suppose that there exist $0 \leq \beta < 4/N$ and two constants A and B such that*

$$G(z) \leq A|z|^2 + B|z|^{\beta+2},$$

for all $z \in \mathbb{C}$. Then, for all $\varphi \in H^1(\mathbb{R}^N)$, we have $T(\varphi) = \infty$, and $\sup_{t \geq 0} \|u(t)\|_{H^1} < \infty$.

Proof. Observe first that, for all $w \in H^1(\mathbb{R}^N)$, we have

$$|V(w)| \leq A \int_{\mathbb{R}^N} |u|^2 + B \int_{\mathbb{R}^N} |u|^{\beta+2}.$$

Applying Theorem 1.3.4, we deduce that

$$|V(w)| \leq A \int_{\mathbb{R}^N} |w|^2 + C \left(\int_{\mathbb{R}^N} |w|^2 \right)^{\frac{\beta}{2}+1-\frac{N\beta}{4}} \left(\int_{\mathbb{R}^N} |\nabla w|^2 \right)^{\frac{N\beta}{4}}. \quad (7.54)$$

Now let $\varphi \in H^1(\mathbb{R}^N)$ and let u be the corresponding solution of (7.4). Let $f(t) = \|\nabla u(t)\|_{L^2}^2$, for $t \in [0, T(\varphi))$. By the conservation of energy, we have

$$f(t) \leq E(\varphi) + 2|V(u(t))|,$$

for all $t \in [0, T(\varphi))$. Applying (7.54), and the conservation of the norm in $L^2(\mathbb{R}^N)$, we deduce that

$$f(t) \leq E(\varphi) + C(\varphi) f(t)^{\frac{N\beta}{4}},$$

for all $t \in [0, T(\varphi))$; hence the result, since $N\beta < 4$. □

Remark 7.5.2. If $\beta = 4/N$, the above inequality becomes

$$f(t) \leq E(\varphi) + C\|\varphi\|_{L^2}^{\frac{4}{N}} f(t),$$

for all $t \in [0, T(\varphi))$. Therefore, if $\|\varphi\|_{L^2}$ is sufficiently small, we again have $T(\varphi) = \infty$.

Proposition 7.5.3. Let g be as in Theorem 7.4.1. There exist $\delta, K > 0$ such that if $\|\varphi\|_{H^1} \leq \delta$, then we have $T(\varphi) = \infty$ and $\sup_{t \geq 0} \|u(t)\|_{H^1} \leq K\|\varphi\|_{H^1}$.

Proof. Observe that

$$G(z) \leq A|z|^2 + B|z|^{\alpha+2},$$

for all $z \in \mathbb{C}$. We may as well assume that $\alpha > 0$. It follows that

$$2|V(w)| \leq C\|w\|_{L^2}^2 + D\|w\|_{H^1}^{\alpha+2}, \qquad (7.55)$$

for all $w \in H^1(\mathbb{R}^N)$. Now let $\varphi \in H^1(\mathbb{R}^N)$ and let u be the corresponding solution of (7.4). Let $f(t) = \|u(t)\|_{H^1}^2$ for $t \in [0, T(\varphi))$. By the conservation laws (ii) and (iii), we have

$$f(t) \leq f(0) + 2|V(\varphi)| + 2|V(u(t))|,$$

for all $t \in [0, T(\varphi))$. Applying (7.55) and conservation of the L^2 norm, and letting $\varepsilon = \alpha/2$, it follows that

$$f(t) \leq f(0) + Cf(0) + Df(0)^{1+\varepsilon} + Cf(0) + Df(t)^{1+\varepsilon},$$

for all $t \in [0, T(\varphi))$. Therefore, if we suppose that $f(0) \leq 1$ then, letting $M = 1 + 2C + D$, we have

$$f(t) \leq Mf(0) + Df(t)^{1+\varepsilon}, \qquad (7.56)$$

for all $t \in [0, T(\varphi))$. Set $\theta(x) = Dx^{1+\varepsilon} - x$ for $x \geq 0$, and let $-\chi = \min \theta < 0$. For all $a \in (0, \chi)$ there exist x_a and y_a with $0 < x_a < y_a$ such that

$$\theta(x_a) + a = \theta(y_a) + a = 0.$$

In addition, we have $a < x_a < a(1+\varepsilon)/\varepsilon$ (see Figure 7.1).

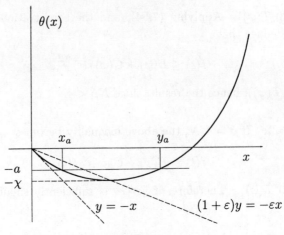

Fig. 7.1

By (7.56), we have
$$\theta(f(t)) + Mf(0) \geq 0,$$
for all $t \in [0, T(\varphi))$. Consequently, if we suppose that $Mf(0) < \chi$, then we have
$$f(t) \in [0, x_{Mf(0)}) \cup (y_{Mf(0)}, \infty).$$
In addition, since f is continuous and $f(0) < x_{Mf(0)}$, we deduce that
$$f(t) \leq x_{Mf(0)} \leq \frac{1+\varepsilon}{\varepsilon} Mf(0),$$
for all $t \in [0, T(\varphi))$. The result follows, with $\delta = \min\{1, \sqrt{\chi/M}\}$ and $K = \sqrt{(1+\varepsilon)M/\varepsilon}$. □

7.6. The non-linear Schrödinger equation in $\mathbb{R}^N$: blow-up in finite time

The blow-up results for the non-linear Schrödinger equation are based on the following proposition.

Proposition 7.6.1. Let $\varphi \in H^1(\mathbb{R}^N)$ and let $u \in C([0, T(\varphi)), H^1(\mathbb{R}^N))$ be the corresponding solution of (7.4). If $|\cdot|\varphi(\cdot) \in L^2(\mathbb{R}^N)$, then we have $|\cdot|u(t,\cdot) \in C([0, T(\varphi)), L^2(\mathbb{R}^N))$. In addition, the function $t \mapsto \int |x|^2 |u(t,x)|^2 \, dx$ belongs to $C^2([0, T(\varphi)))$, and we have

$$\frac{d}{dt}\int_{\mathbb{R}^N}|x|^2|u(t,x)|^2\,dx = -4\langle iu_r, ru\rangle = 4\,\text{Im}\int_{\mathbb{R}^N} r\bar{u}u_r\,dx, \tag{7.57}$$

$$\frac{d^2}{dt^2}\int_{\mathbb{R}^N}|x|^2|u(t,x)|^2\,dx = 16E(u(t))$$
$$-4N\int_{\mathbb{R}^N}g(|u(t)|)|u(t)|\,dx + 8(N+2)\int_{\mathbb{R}^N}G(u(t))\,dx, \tag{7.58}$$

for all $t \in [0, T(\varphi))$.

Remark. Identity (7.58) is very important. It is useful not only for establishing blow-up results, but also for studying the global behaviour of solutions (Section 7.7).

The proof of (7.57) is rather simple. Formally, it suffices to multiply the equation by $r^2 u$. However, this is not correct because of the term r^2. We overcome this difficulty multiplying the equation by $e^{-\varepsilon r^2} r^2 u$ and passing to the limit as $\varepsilon \downarrow 0$ (Lemma 7.6.2). The proof of (7.58) is much more complicated. Formally, we would multiply the equation by ru_r though this is not possible for two reasons: firstly, because of the term in r; secondly, u is only in $H^1(\mathbb{R}^N)$ so the equation makes sense in $H^{-1}(\mathbb{R}^N)$, but u_r is only in $L^2(\mathbb{R}^N)$. Therefore, we cannot multiply the equation by u_r (for this we need u_r in $H^1(\mathbb{R}^N)$). To show (7.58), we then proceed as follows: we approximate φ by a sequence of functions φ_n of $H^2(\mathbb{R}^N)$, and we denote by u_n the corresponding solutions. We multiply the equation satisfied by u_n by $e^{-\frac{\varepsilon r^2}{2}}r\partial_r u_n$ and then we let $\varepsilon \downarrow 0$. We obtain (7.58) for the solution u_n (Lemma 7.6.3) and then we let $n \to \infty$.

Lemma 7.6.2. Let $\varphi \in H^1(\mathbb{R}^N)$ and let $u \in C([0, T(\varphi)), H^1(\mathbb{R}^N))$ be the corresponding solution of (7.4). If $|\cdot|\varphi(\cdot) \in L^2(\mathbb{R}^N)$, we have $|\cdot|u(t,\cdot) \in C([0, T(\varphi)), L^2(\mathbb{R}^N))$. In addition, the function $t \mapsto \int |x|^2 |u(t,x)|^2\,dx$ is in $C^1([0, T(\varphi)))$, and identity (7.57) holds.

Proof. For $\varepsilon > 0$, we define the functions θ_ε and ρ_ε by

$$\theta_\varepsilon(r) = re^{-\varepsilon\frac{r^2}{2}};$$
$$\rho_\varepsilon(r) = r^2 e^{-\varepsilon r^2}.$$

We set

$$f_\varepsilon(t) = \int_{\mathbb{R}^N}|\theta_\varepsilon u|^2\,dx,$$

for $t \in [0, T(\varphi))$. It is clear that $f_\varepsilon \in C^1([0, T(\varphi)))$, and that, for all $t \in [0, T(\varphi))$, we have

$$\begin{aligned} f'_\varepsilon(t) &= 2\langle \theta_\varepsilon u, \theta_\varepsilon u_t \rangle_{H^1, H^{-1}} = 2\langle \rho_\varepsilon u, u_t \rangle = 2\langle \rho_\varepsilon u, i\Delta u + ig(u) \rangle \\ &= 2\langle \rho_\varepsilon u, i\Delta u \rangle = -2\langle \rho_\varepsilon \nabla u, i\nabla u \rangle - 2\langle \nabla \rho_\varepsilon u, i\nabla u \rangle \\ &= -2\langle \nabla \rho_\varepsilon u, i\nabla u \rangle = -2\langle \rho'_\varepsilon u, iu_r \rangle. \end{aligned} \quad (7.59)$$

We easily verify that $|\rho'_\varepsilon(r)| \leq C\theta_\varepsilon(r)$, and so we obtain

$$f'_\varepsilon(t) \leq 2C\|u(t)\|_{H^1} f_\varepsilon(t)^{1/2}.$$

It follows immediately that $f_\varepsilon(t)$ is bounded as $\varepsilon \downarrow 0$, uniformly on $[0, T]$, for all $T < T(\varphi)$. By monotone convergence, we deduce that $\int |x|^2 |u(t,x)|^2 \, dx < \infty$ for all $t \in [0, T(\varphi))$, and that $\| |\cdot| u(t,\cdot) \|_{L^2}$ is bounded on any interval $[0, T]$, with $T < T(\varphi)$. In particular, $t \mapsto |\cdot| u(t, \cdot)$ is weakly continuous from $[0, T(\varphi))$ to $L^2(\mathbb{R}^N)$. Integrating (7.59) between 0 and $t \in [0, T(\varphi))$, we obtain

$$f_\varepsilon(t) = f_\varepsilon(0) - 4\int_0^t \langle (1 - \varepsilon r^2) e^{-\varepsilon \frac{r^2}{2}}, iu_r \rangle,$$

for all $t \in [0, T(\varphi))$. Letting $\varepsilon \downarrow 0$, it follows that

$$\int_{\mathbb{R}^N} |xu(t)|^2 \, dx = \int_{\mathbb{R}^N} |x\varphi|^2 \, dx - 4\int_0^t \langle ru, iu_r \rangle,$$

for all $t \in [0, T(\varphi))$. We deduce (7.57). In addition, the function

$$t \mapsto \| |\cdot| u(t, \cdot) \|_{L^2}$$

is continuous, and then $|\cdot| u(t, \cdot) \in C([0, T(\varphi)), L^2(\mathbb{R}^N))$. □

Lemma 7.6.3. Let $\varphi \in H^2(\mathbb{R}^N)$ and let $u \in C([0, T(\varphi)), H^2(\mathbb{R}^N))$ be the corresponding solution of (7.4). If $|\cdot| \varphi(\cdot) \in L^2(\mathbb{R}^N)$, we have $|\cdot| u(t, \cdot) \in C([0, T(\varphi)), L^2(\mathbb{R}^N))$. In addition, the function $t \mapsto \int |x|^2 |u(t,x)|^2 \, dx$ is in $C^2([0, T(\varphi)))$, and (7.58) holds.

Proof. We proceed in five steps.

Step 1. Let $\theta \in \mathcal{S}(\mathbb{R}^N)$ be a radially symmetric real-valued function. Let $T > 0$ and $u \in C([0, T], H^1(\mathbb{R}^N)) \cap C^1([0, T], L^2(\mathbb{R}^N))$. Set $h(t) = \langle r\theta u, iu_r \rangle$, for $t \in [0, T]$. Then $h \in C^1([0, T])$ and

$$h'(t) = \langle u_t, i(2\theta r u_r + (N\theta + r\theta')u) \rangle, \quad (7.60)$$

for all $t \in [0,T]$. Indeed, suppose first that $u \in C^1([0,T], H^1(\mathbb{R}^N))$. In that case, it is clear that $h \in C^1([0,T])$ and we have

$$h'(t) = \langle r\theta u_t, iu_r\rangle + \langle r\theta u, iu_{tr}\rangle = \langle u_t, ir\theta u_r\rangle + \langle \theta u, ix \cdot \nabla u_t\rangle. \quad (7.61)$$

Observe that

$$\langle \theta u, ix \cdot \nabla u_t\rangle = \operatorname{Im}\int_{\mathbb{R}^N} \theta x \cdot \nabla \overline{u_t}\, dx,$$

and that

$$\theta x \cdot \nabla \overline{u_t} = \nabla \cdot (x\theta u \overline{u_t}) - N\theta u \overline{u_t} - \theta \overline{u_t} x \cdot \nabla u - r\theta' u \overline{u_t}.$$

We then have

$$\langle \theta u, ix \cdot \nabla u_t\rangle = \langle u_t, i(\theta x \cdot \nabla u + (N\theta + r\theta')u)\rangle = \langle u_t, i(\theta r u_r + (N\theta + r\theta')u)\rangle.$$

Applying (7.61), we deduce (7.60), and then

$$h(t) = h(0) + \int_0^t \langle u_t, i(2\theta r u_r + (N\theta + \theta')u)\rangle, \quad (7.62)$$

for all $t \in [0,T]$. By density, we obtain (7.62) as well as (7.60) for $u \in C([0,T], H^1(\mathbb{R}^N)) \cap C^1([0,T], L^2(\mathbb{R}^N))$.

Step 2. Let u be as in the statement of the lemma and set $h(t) = \langle r\theta u, iu_r\rangle$, for $t \in [0,T]$. Then, by Step 1, we have $h \in C^1([0, T(\varphi)))$ and

$$h'(t) = \langle \Delta u + g(u), 2\theta r u_r + (N\theta + r\theta')u\rangle, \quad (7.63)$$

for all $t \in [0,T]$.

Step 3. Let $w \in H^2(\mathbb{R}^N)$ be such that $|\cdot|w(\cdot) \in L^2(\mathbb{R}^N)$. Then we have

$$\langle \Delta w + g(w), 2\theta r w_r + (N\theta + r\theta')w\rangle = -2\int_{\mathbb{R}^N} \theta|\nabla w|^2$$
$$+ N\int_{\mathbb{R}^N} \theta(g(|w|)|w| - 2G(w))$$
$$+ \int_{\mathbb{R}^N} r\theta'(g(|w|)|w| - 2G(w) - 2|w_r|^2)$$
$$- \operatorname{Re}\int_{\mathbb{R}^N} r((N+1)\theta' + r\theta'')w_r \overline{w}. \quad (7.64)$$

Note first that, by density, it suffices to consider the case $w \in \mathcal{D}(\mathbb{R}^N)$. In that case, we have

$$\langle g(w), (N\theta + r\theta')w\rangle = \int (N\theta + r\theta')g(|w|)|w|; \quad (7.65)$$

118 *The Schrödinger equation*

$$\langle g(w), 2\theta r w_r \rangle = \operatorname{Re} \int_{\mathbb{R}^N} 2\theta r g(w) \overline{w_r} = \int_{\mathbb{R}^N} 2\theta r G(w)_r.$$

But $2\theta r G(w)_r = 2\theta x \cdot \nabla G(w) = \nabla \cdot (x 2\theta G(w)) - 2(N\theta + r\theta') G(w)$. Consequently,

$$\langle g(w), 2\theta_r w_r \rangle = -2 \int (N\theta + r\theta') G(w). \tag{7.66}$$

On the other hand, we have

$$\langle \Delta w, (N\theta + r\theta') w \rangle = -\int_{\mathbb{R}^N} (N\theta + r\theta') |\nabla w|^2$$
$$- \operatorname{Re} \int_{\mathbb{R}^N} r((N+1)\theta' + r\theta'') w_r \overline{w}; \tag{7.67}$$

$$\langle \Delta w, 2\theta r w_r \rangle = -\langle 2\theta \nabla u, \nabla(r u_r) \rangle - 2 \int_{\mathbb{R}^N} r\theta' |w_r|^2. \tag{7.68}$$

Note also that the following identity holds:

$$\operatorname{Re}(2\theta \nabla w \cdot \nabla(r \overline{w_r})) = \operatorname{Re}(2\theta \nabla w \cdot \nabla(x \cdot \nabla \overline{w}))$$
$$= ((2-N)\theta - r\theta') |\nabla w|^2 + \nabla \cdot (x \theta |\nabla w|^2).$$

From (7.68), we then deduce that

$$\langle \Delta w, 2\theta_r w_r \rangle = \int_{\mathbb{R}^N} ((N-2)\theta + r\theta') |\nabla w|^2 - 2 \int_{\mathbb{R}^N} r\theta' |w_r|^2. \tag{7.69}$$

Applying (7.65), (7.66), (7.67), and (7.69), we obtain (7.64).

Step 4. Let u be as in the statement of the lemma and set $h(t) = \langle r\theta u, iu_r \rangle$, for $t \in [0, T(\varphi))$. Then, by Steps 2 and 3, we have

$$h'(t) = -2 \int_{\mathbb{R}^N} \theta |\nabla u|^2 + N \int_{\mathbb{R}^N} \theta(g(|u|)|u| - 2G(u))$$
$$+ \int_{\mathbb{R}^N} r\theta'(g(|u|)|u| - 2G(u) - 2|u_r|^2) \tag{7.70}$$
$$- \operatorname{Re} \int_{\mathbb{R}^N} r((N-1)\theta' + r\theta'') u_r \overline{u},$$

for all $t \in [0, T(\varphi))$.

Step 5. Take $\theta_\varepsilon = e^{-\varepsilon r^2}$, for $\varepsilon > 0$. We have $|\theta_\varepsilon| \le 1$ and $\theta_\varepsilon \to 1$ as $\varepsilon \downarrow 0$. We easily verify that $|r\theta'_\varepsilon| \le C$, $r\theta'_\varepsilon \longrightarrow 0$ as $\varepsilon \downarrow 0$, and $|r[(N+1)\theta'_\varepsilon + r\theta''_\varepsilon]| \le C$ and that $|r[(N+1)\theta'_\varepsilon + r\theta''_\varepsilon]| \longrightarrow 0$ as $\varepsilon \downarrow 0$. We deduce immediately from (7.70) that

$$\langle r\theta_\varepsilon u, iu_r \rangle = \langle r\theta_\varepsilon \varphi, i\varphi_r \rangle + \int_0^t f_\varepsilon(s) \, ds, \tag{7.71}$$

where f_ε is a bounded function on $[0,T]$ for all $T \in [0, T(\varphi))$ and is such that

$$f_\varepsilon(t) \longrightarrow -2\int_{\mathbb{R}^N} |\nabla u|^2 + N\int_{\mathbb{R}^N} (g(|u|)|u| - 2G(u)) \quad \text{as } \varepsilon \downarrow 0. \tag{7.72}$$

Observe that, due to conservation of energy, we have

$$-2\int_{\mathbb{R}^N} |\nabla u|^2 + N\int_{\mathbb{R}^N} (g(|u|)|u| - 2G(u))$$
$$= -4E(u(t)) + N\int_{\mathbb{R}^N} g(|u|)|u| - (2N+4)\int G(u).$$

Letting $\varepsilon \downarrow 0$ in (7.71), we obtain the following:

$$\langle ru, iu_r \rangle = \langle r\varphi, i\varphi_r \rangle - \int_0^t \Big(4E(u(t)) - N\int_{\mathbb{R}^N} g(|u|)|u| \\ + (2N+4)\int_{\mathbb{R}^N} G(u) \Big). \tag{7.73}$$

Now, (7.58) follows readily from (7.57) and (7.73). □

Proof of Proposition 7.6.1. Applying Lemma 7.6.2, it remains to verify that the function

$$t \mapsto \int_{\mathbb{R}^N} |x|^2 |u(t,x)|^2 \, dx$$

is in $C^2([0, T(\varphi)))$, and that we have (7.58). Let $T < T(\varphi)$ and let $(\varphi_m)_{m \geq 0}$ be a sequence of functions in $H^2(\mathbb{R}^N)$, such that $\varphi_m \to \varphi$ in $H^1(\mathbb{R}^N)$, as $m \to \infty$. Denote the corresponding solutions of (7.4) by u_m. We know (Theorem 7.4.1) that, for m sufficiently large, we have $T(\varphi_m) > T$ and $u_m \to u$ in $C([0,T], H^1(\mathbb{R}^N))$, as $m \to \infty$. We write identity (7.73) for u_m and $t \in [0,T]$ and we let $m \to \infty$. We deduce that u satisfies (7.73); hence the result. □

Theorem 7.6.4. *Let g be as in Theorem 7.4.1, and suppose further that*

$$sg(s) \geq \left(2 + \frac{4}{N}\right) G(s), \tag{7.74}$$

for all $s \geq 0$. Then if $\varphi \in H^1(\mathbb{R}^N)$ is such that $|\cdot| \, |\varphi(\cdot)| \in L^2(\mathbb{R}^N)$ and $E(\varphi) < 0$, we have $T(\varphi) < \infty$.

Proof. Let φ be as above. Let u be the corresponding solution of (7.4) and set

$$f(t) = \int_{\mathbb{R}^N} |x|^2 |u(t,x)|^2 \, dx,$$

for $t \in [0, T(\varphi))$. It follows from (7.58), (7.74), and the conservation of energy that
$$f''(t) \leq 16E(\varphi), \qquad (7.75)$$
for all $t \in [0, T(\varphi))$. From (7.75), we deduce that
$$f(t) \leq f(0) + tf'(0) + 8t^2 E(\varphi),$$
for all $t \in [0, T(\varphi))$; this implies that $T(\varphi) < \infty$, since $f(t) \geq 0$ and $E(\varphi) < 0$. □

Remark 7.6.5. If there exists $x > 0$ such that $G(x) > 0$, (7.74) implies that $G(s) \geq (s/x)^{2+4/N} G(x)$ for $s \geq x$. In particular, if we take $\varphi \in H^1(\mathbb{R}^N)$, then $E(k\varphi) < 0$ for k large enough; and so if $|\cdot| \varphi(\cdot) \in L^2(\mathbb{R}^N)$, then $T(k\varphi) < \infty$.

7.7. A remark concerning behaviour at infinity

Identities (7.57) and (7.58) allow us to prove directly the pseudo-conformal conservation law, which provides information about the behaviour at infinity in time of the solutions, in some cases (see §7.8 below). The following proposition is related to this conservation law.

Proposition 7.7.1. Let $\varphi \in H^1(\mathbb{R}^N)$ be such that $|\cdot| \varphi(\cdot) \in L^2(\mathbb{R}^N)$ and let u be the corresponding solution of (7.4). Set
$$f(t) = \int_{\mathbb{R}^N} |(x + 2it\nabla) u(t, x)|^2 \, dx - 8t^2 \int_{\mathbb{R}^N} G(u(t, x)) \, dx, \qquad (7.76)$$
for all $t \in [0, T(\varphi))$. Then $f \in C^1([0, T(\varphi)))$ and
$$f'(t) = 4t \int_{\mathbb{R}^N} (Ng(|u|)|u| - 2(N+2)G(u)), \qquad (7.77)$$
for all $t \in [0, T(\varphi))$.

Proof. Developing the right-hand side of (7.76), we obtain
$$f(t) = \|xu\|_{L^2}^2 + 4t\langle ru, iu_r\rangle + 4t^2 \|\nabla u\|_{L^2}^2 - 8t^2 V(u)$$
$$= \|xu\|_{L^2}^2 + 4t\langle ru, iu_r\rangle + 8t^2 E(\varphi).$$

It follows immediately from Proposition 7.6.1 that $f \in C^1([0, T(\varphi)))$ and that identity (7.77) holds. □

Remark 7.7.2. If $g(s) = \lambda s^{1+4/N}$, (7.77) means $f'(t) = 0$, and then we have $f(t) = f(0) = \int |x\varphi|^2$.

Remark 7.7.3. Let u be as in the statement of Proposition 7.7.1, and set

$$v(t,x) = e^{-i\frac{|x|^2}{4t}} u(t,x),$$

for $x \in \mathbb{R}^N$ and $t \in [0, T(\varphi))$. It is clear that

$$|(x + 2it\nabla)u(t)|^2 = 4t^2|\nabla v(t)|^2 \geq 4t^2|\nabla|u(t)||^2;$$

and, consequently,

$$f(t) = 8t^2 E(v(t)) \geq 8t^2 E(|u(t)|). \tag{7.78}$$

7.8. Application to a model case

We choose $g(s) = a|s|^\alpha s$, with $\alpha > 0$, $(N-2)\alpha < 4$, and $a \neq 0$. We consider $\varphi \in H^1(\mathbb{R}^N)$ and we denote by u the corresponding maximal solution of (7.4), which exists by Theorem 7.4.1. We then have the following results.

If $a < 0$, then $T(\varphi) = \infty$ and u is bounded in $H^1(\mathbb{R}^N)$ (Proposition 7.5.1).

If $a > 0$ and if $\alpha < 4/N$, then $T(\varphi) = \infty$ and u is bounded in $H^1(\mathbb{R}^N)$ (Proposition 7.5.1).

If $a > 0$ and if $\alpha = 4/N$, then $T(\varphi) < \infty$ for some initial data φ (Theorem 7.6.4 and Remark 7.6.5); on the other hand, if $\|\varphi\|_{L^2}$ is sufficiently small then $T(\varphi) = \infty$ and u is bounded in $H^1(\mathbb{R}^N)$ (Remark 7.5.2).

If $a > 0$ and if $\alpha > 4/N$, we have $T(\varphi) < \infty$ for some special initial data φ (Theorem 7.6.4 and Remark 7.6.5); on the other hand, if $\|\varphi\|_{H^1}$ is small enough, then $T(\varphi) = \infty$ and u is bounded in $H^1(\mathbb{R}^N)$ (Proposition 7.5.3).

Observe that in the case in which $a < 0$ and if $|\cdot|\varphi(\cdot) \in L^2(\mathbb{R}^N)$, (7.77) shows that the solutions converge to 0 as $t \to \infty$, in certain spaces $L^p(\mathbb{R}^N)$. Indeed, it follows from (7.77) that

$$f'(t) = -4(4 - N\alpha)t \frac{a}{\alpha+2} \int_{\mathbb{R}^N} |u(t)|^{\alpha+2} = 4(4-N\alpha)tV(u(t)), \tag{7.79}$$

for all $t \geq 0$. In particular, if $\alpha \geq \frac{4}{N}$, we have $f'(t) \leq 0$, and so, with (7.78), it follows that

$$8t^2 E(|u(t)|) \leq \int |x\varphi|^2,$$

for all $t > 0$. If we apply Theorem 1.3.4, we obtain, in particular,

$$\|u(t)\|_{L^{\alpha+2}} \leq C \|\nabla |u(t)|\|_{L^2}^{N(\frac{1}{2}-\frac{1}{\alpha+2})} \|\varphi\|_{L^2}^{1-N(\frac{1}{2}-\frac{1}{\alpha+2})} \leq C t^{-\frac{N\alpha}{2(\alpha+2)}},$$

for all $t \geq 0$. Observe that we obtain the same negative exponent of t as for the linear equation (Proposition 3.5.14).

If $\alpha < 4/N$, it follows from (7.79) and (7.78) that

$$f'(t) \leq 4(4 - N\alpha)tE(|u(t)|) \leq \left(2 - \frac{N\alpha}{2}\right)\frac{1}{t}f(t),$$

for all $t > 0$. Therefore, $t^{\frac{N\alpha-4}{2}}f(t)$ is non-decreasing and, (by (7.78)), we have

$$8t^{\frac{N\alpha}{2}}E(|u(t)|) \leq f(1),$$

for all $t \geq 1$. Consequently, we again obtain

$$\|u(t)\|_{L^{\alpha+2}} \leq Ct^{-\frac{N\alpha}{2(\alpha+2)}},$$

i.e. the same negative exponenet of t as for the linear equation.

Notes. For $\Omega \neq \mathbb{R}^N$, $N \geq 2$, a few results are known. See Brezis and Gallouet [1], Kavian [1], Y. Tsutsumi [2, 3], and Yao [1]. Note also that Cazenave and Haraux's results [1] apply in any open subset $\Omega \subset \mathbb{R}^N$.

For §7.3, see Yajima [1], and Strichartz [1]. A regularizing effect of $H^s(\mathbb{R}^N)$-type also exists; see Constantin and Saut [1–3], Sjölin [1], and Vega [1]. The problem of local existence (§7.4) has been studied extensively. Our presentation is based on Kato [3] and Cazenave and Weissler [1, 3]. See also Baillon, Cazenave, and Figueira [1], Cazenave [2], Ginibre and Velo [1, 2, 5], Hayashi [1], Hayashi and Tsutsumi [1], Lin and Strauss [1], and Weinstein [1]. The Cauchy problem has also been studied in $H^s(\mathbb{R}^N)$ ($s \neq 1$); see Cazenave and Weissler [2, 4], Ginibre and Velo [4], and Y. Tsutsumi [4]. There exists a regularizing effect for the non-linear equation; see Hayashi, Nakamitsu, and Tsutsumi [1, 2], Hayashi and Ozawa [2], and Kato [3].

Various kinds of non-linearities (possibly non-local) have been considered. See, for example, Baillon, Cazenave, and Figueira [2], Baillon and Chadam [1], Cazenave [2], Cazenave and Haraux [1], Cazenave and Weissler [1], Chadam and Glassey [1], Dias and Figueira [1], Ginibre and Velo [3], Klainerman and Ponce [1], Lange [1], and Schochet and Weinstein [1]. For more about blow-up in finite time, consult M. Tsutsumi [1], Glassey [2], and Weinstein [4].

For some non-linearities, there exist solutions of the form $u(t,x) = e^{i\omega t}\varphi(x)$. These solutions are called stationary states. See, for example, Berestycki, Gallouet and Kavian [1], Berestycki and Lions [1], Berestycki, Lions, and Peletier [1], and Jones and Küpper [1].

The behaviour at infinity of solutions is rather well known in the repulsive case, if the solutions behave asymptotically as the solutions of the linear equation. See Ginibre and Velo [1, 2, 7, 8, 10], Hayashi [2], Hayashi and Ozawa [1],

Lin and Strauss [1], Reed and Simon [1], and Y. Tsutsumi [1]. In the attractive case, we only know how to study the stability of some stationary states. See Berestycki and Cazenave [1], Blanchard, Stubbe, and Vazquez [1], Cazenave [3], Cazenave and Lions [1], Grillakis, Shatah, and Strauss [1], Jones [1], Shatah and Strauss [1], and Weinstein [2, 3]. For more references concerning these questions, consult Cazenave [4].

8
Bounds on global solutions

The study of the behaviour at infinity of global solutions is one of the most important problems in the study of non-linear evolution equations. The problem can be formulated as follows. If $u(t)$, $0 \le t < \infty$ is a solution of an equation of the form

$$u'(t) = Au(t) + F(t, u(t)),$$

for $t \ge 0$, how does $u(t)$ behave as $t \to \infty$? The results concerning the behaviour at infinity of global solutions are, in general, based on compactness properties of $\bigcup_{t \ge 0} \{u(t)\}$, and in particular on bounds for $\bigcup_{t \ge 0} \{u(t)\}$ (see Chapter 9). In some cases, the solutions are bounded by construction (see §5.3, §6.3, §7.5). On the other hand, even for linear equations, global solutions may be not bounded. For example, $u(t,x) = e^{\pi t} \sin(\pi x)$ is a solution of

$$\begin{cases} u_t = u_{xx} + 2\pi^2 u, & \text{in } \Omega = (0,1); \\ u = 0, & \text{in } \partial\Omega; \end{cases}$$

and u is bounded in no function space as $t \to \infty$.

The available methods allow us to study the behaviour of the solutions in the following two situations:

- Semilinear autonomous equations (see §8.1, §8.2, §5.3, §6.3, and §7.5).
- Semilinear non-autonomous equations, with dissipation (§8.4) or with repulsive interaction (see §8.3).

8.1. The heat equation

We use the notation of Chapter 5. In particular, Ω is a bounded open subset of $\mathbb{R}^N$ with Lipschitz continuous boundary, $X = C_0(\Omega)$ and $(T(t))_{t \ge 0}$ denotes the semigroup associated with the heat equation. g is a locally Lipschitz continuous function $\mathbb{R} \to \mathbb{R}$ such that $g(0) = 0$, and we consider G and E to be defined as for Proposition 5.4.4.

In the following subsection, we gather some preliminary results that will be useful in establishing the main result.

8.1.1. A singular Gronwall's lemma: application to the heat equation

Lemma 8.1.1. Let $T > 0$, $A \geq 0$, $0 \leq \beta \leq \alpha$, and $1 \leq \gamma \leq \infty$. Let $\varphi \in C([0,T])$, $\varphi \geq 0$ and $f \in L^\gamma(0,T)$, $f \geq 0$. Suppose that $\alpha + 1/\delta < 1$ and that

$$\varphi(t) \leq At^{-\alpha} + \int_0^t \frac{1}{(t-s)^\beta} f(s)\varphi(s)\,ds,$$

for all $t \in (0,T)$. Then $\varphi(t) \leq CAt^{-\alpha}$, for all $t \in (0,T)$, where C depends only on T, α, β, γ, and $\|f\|_{L^\gamma(0,T)}$.

Proof. Set $\psi(t) = t^\alpha \varphi(t)$ for all $t \in [0,T]$, and let $\theta(t) = \sup_{0 \leq s \leq t} \psi(s)$ for $t \in [0,T]$. It is clear that $\theta \in C([0,T])$ and that we have

$$\psi(t) \leq A + t^\alpha \int_0^t \frac{1}{(t-s)^\beta} \frac{1}{s^\alpha} f(s)\theta(s)\,ds,$$

for all $t \in (0,T)$. We have $t^{-\beta} \in L^{\gamma'}(0,T)$, and so there exists $\varepsilon \in (0,1)$ such that

$$(1-\varepsilon)^{-\alpha} \|t^{-\beta}\|_{L^{\gamma'}(0,\varepsilon T)} \|f\|_{L^\gamma(0,T)} \leq 1/2.$$

It follows that

$$\psi(t) \leq A + t^\alpha \theta(t) \int_{(1-\varepsilon)t}^t \frac{1}{(t-s)^\beta} \frac{1}{s^\alpha} f(s)\,ds$$
$$+ t^\alpha \int_0^{(1-\varepsilon)t} \frac{1}{(t-s)^\beta} \frac{1}{s^\alpha} f(s)\theta(s)\,ds$$
$$\leq A + \frac{1}{2}\theta(t) + \frac{1}{\varepsilon^\beta} t^{\alpha-\beta} \int_0^{(1-\varepsilon)t} \frac{1}{s^\alpha} f(s)\theta(s)\,ds.$$

We deduce immediately that

$$\theta(t) \leq A + \frac{1}{2}\theta(t) + \frac{1}{\varepsilon^\beta} T^{\alpha-\beta} \int_0^T \frac{1}{s^\alpha} f(s)\theta(s)\,ds;$$

and we obtain the conclusion by applying Lemma 4.2.1, since $s^{-\alpha} f(s) \in L^1(0,T)$. □

We will also make use of the following comparison lemma, which generalizes Proposition 5.3.1.

Lemma 8.1.2. Let $T > 0$, $p \ge 2N/(N+2)$ ($p \ge 1$ if $N = 1$, $p > 1$ if $N = 2$), $\varphi \in X$, $u \in C([0,T], X) \cap L^2((0,T), H_0^1(\Omega)) \cap W^{1,2}((0,T), H^{-1})$ and $f \in L^1((0,T), L^p(\Omega))$ such that

$$\begin{cases} u_t = \Delta u + f & \text{in } H^{-1}(\Omega), \text{ almost everywhere in } (0,T); \\ u(0) = \varphi. \end{cases}$$

Suppose further that there exists a constant C such that

$$f^-(t,x) \le C|u(t,x)| \quad \text{almost everywhere in } \Omega,$$

for almost every $t \in (0,T)$. Then if $\varphi \ge 0$ on Ω, we have $u(t) \ge 0$ on Ω, for all $t \in [0,T]$.

Proof. Note first that the hypotheses imply that $f \in L^1((0,T), H^{-1}(\Omega))$, and so the equation makes sense in $H^{-1}(\Omega)$. Now, the proof of Proposition 5.3.1 can be adapted immediately, since

$$\frac{d}{dt} \int_\Omega u^-(t)^2 \, dx = 2\langle u_t(t), u^-(t) \rangle_{H^{-1}, H_0^1} = -2 \int_\Omega |\nabla u^-(t)|^2 - 2 \int_\Omega f(t) u^-(t),$$

almost everywhere in Ω. □

Lemma 8.1.3. Let $T > 0$, $\sigma, \gamma \ge 1$ be such that $N/(2\sigma) + 1/\gamma < 1$. Let $\varphi \in X$, $u \in C([0,T], X)$ and $f \in L^\infty((0,T), L^\sigma(\Omega))$ be such that

$$u(t) = \mathcal{T}(t)\varphi + \int_0^t \mathcal{T}(t-s)\{f(s)u(s)\} \, ds,$$

for all $t \in [0,T]$. Then for all $\varepsilon \in (0,T]$, we have $\sup_{\varepsilon \le t \le T} \|u(t)\|_{L^\infty} \le K$, where K depends only on $\varepsilon, \sigma, \gamma, \|\varphi\|_{L^1}$ and $\|f^+\|_{L^\gamma((0,T), L^\sigma)}$.

Proof. We proceed in two steps. Note first that the hypotheses imply $fu \in L^\infty((0,T), L^\sigma(\Omega))$, so the above integral makes sense, since by (3.34) $\mathcal{T}(t)$ can be extended to an operator in $\mathcal{L}(L^p, L^p)$ for all $p \in [1, \infty)$. On the other hand, invoking Lemma 8.1.2, we may restrict ourselves to the case in which $\varphi \ge 0$, and so $u \ge 0$ (see the proof of Proposition 5.3.3).

Step 1. Let $1 \le \rho \le r \le \infty$ be such that

$$\frac{N}{2}\left(\frac{1}{\rho} - \frac{1}{r}\right) < 1 \quad \text{and} \quad \frac{1}{r} + \frac{1}{\sigma} \le 1.$$

Let us show that, under the hypotheses of the lemma, we have

$$\sup_{\varepsilon \le t \le T} \|u(t)\|_{L^r} \le C,$$

where C is a constant that depends only on $\|f^+\|_{L^\gamma((0,T),L^\sigma)}$, $\varepsilon, \sigma, \gamma, r, \rho, T$, and $\|\varphi\|_{L^r}$. Indeed, by Hölder's inequality, (3.34) and Lemma 3.5.9, we have

$$\|u(t)\|_{L^r} \le t^{-\frac{N}{2}(\frac{1}{\rho}-\frac{1}{r})}\|\varphi\|_{L^\rho} + \int_0^t \frac{1}{(t-s)^{\frac{N}{2\sigma}}} \|f(s)^+\|_{L^\sigma} \|u(s)\|_{L^r}\, ds,$$

for all $t \in (0,T]$. We conclude by applying Lemma 8.1.1. □

Step 2. Let m be an integer such that $m < \sigma \le m+1$. Let $\varepsilon > 0$ and $\delta = \varepsilon/(m+1)$. Applying Step 1 with $\rho = 1$ and $r = \sigma/(\sigma-1)$, we conclude that $u(\delta)$ is bounded in $L^{\sigma/(\sigma-1)}(\Omega)$, with respect to the above parameters. We iterate this argument, translating the time of δ at each Step and taking successively $\rho = \sigma/(\sigma-j)$ and $r = \sigma/(\sigma-j-1)$, for $1 \le j \le m-1$. We obtain that $u(m\delta)$ is bounded in $L^{\sigma/(\sigma-m)}(\Omega)$. We conclude by applying Step 1 again with $\rho = \sigma/(\sigma-m)$ and $r = \infty$, to find that u is bounded in $L^\infty(\Omega)$ on $[\varepsilon, T]$, with respect to the above parameters. □

Corollary 8.1.4. *Let $\sigma \ge 1$ be such that $\sigma > N/2$, $\varphi \in X$, $u \in C([0,\infty), X)$ and let $f \in C([0,\infty), L^\sigma(\Omega))$ be such that*

$$u(t) = T(t)\varphi + \int_0^t T(t-s)\{f(s)u(s)\}\, ds,$$

for all $t \ge 0$. Suppose that $\sup_{t \ge 0} \|u(t)\|_{L^1} < \infty$ and that $\sup_{t \ge 0} \|f(t)^+\|_{L^\sigma} < \infty$. Then, for all $\varepsilon > 0$, we have $\sup_{t \ge \varepsilon} \|u(t)\|_{L^\infty} < \infty$.

Proof. For all $s \ge 0$, we apply Lemma 8.1.3 with $T = 1 + \varepsilon$ and $\varphi = u(s)$. We obtain in particular that $u(s+\varepsilon)$ is estimated in $L^\infty(\Omega)$-norm, only in terms of $\varepsilon, s, \sup_{t \ge 0} \|u(t)\|_{L^1}$ and $\sup_{t \ge 0} \|f(t)^+\|_{L^\sigma}$; hence the result, since $s \ge 0$ is arbitrary. □

Corollary 8.1.5. *Suppose that g satisfies*

$$xg(x) \le C(x^2 + |x|^p), \quad \forall x \in \mathbb{R}, \tag{8.1}$$

where $p \ge 2$ and $(N-2)p < 2N$. Then, for all M, there exist $t(M) > 0$ and $K(M) < \infty$ with the following properties: if $\varphi \in X \cap H_0^1(\Omega)$ is such that $\|\varphi\|_{H^1} \le M$ and if u denotes the corresponding maximal solution of (5.4) (see Theorem 5.2.1), then $T(\varphi) \geq t(M)$ and $\|u(t)\|_{L^\infty} \leq K(M)$ for all $t \in [t(M)/2, t(M)]$.

Proof. Applying Proposition 5.3.1 and arguing as in the proof of Proposition 5.3.3, we may restrict ourselves to the case in which $\varphi \geq 0$, and so $u \geq 0$. From (8.1), it follows that

$$\|g(u(t))^+\|_{L^{\frac{p}{p-1}}} \leq A\left(1 + \|u(t)\|_{L^p}\right)^{p-1}, \tag{8.2}$$

for all $t \in [0, T(\varphi))$. If we set $\beta = N(p-2)/(2p) \in [0,1)$, we deduce from (3.34), (8.2), and Lemma 3.5.9 that

$$\|u(t)\|_{L^p} \leq \|\varphi\|_{L^p} + A \int_0^t \frac{1}{(t-s)^\beta} \left(1 + \|u(s)\|_{L^p}\right)^{p-1} ds,$$

for all $t \in [0, T(\varphi))$. But, using Sobolev's inequalities, we have $\|\varphi\|_{L^p} \leq CM$. Setting $f(t) = 1 + \sup_{0 \leq s \leq t} \|u(s)\|_{L^p}$, we then obtain

$$f(t) \leq (1 + CM) + \frac{At^{1-\beta}}{1-\beta} f(t)^{p-1}, \tag{8.3}$$

for all $t \in [0, T(\varphi))$. Set

$$\tau(M) = \left(\frac{(1-\beta)(1+CM)^{2-p}}{2^p A}\right)^{\frac{1}{1-\beta}}.$$

We easily deduce from (8.3) that $f(t) \leq 2(1+CM)$ for $0 \leq t < \min\{T(\varphi), t(M)\}$. Applying (8.1) we conclude immediately that

$$\left\|\left(\frac{g(u(t))}{u(t)}\right)^+\right\|_{L^{\frac{p}{p-2}}} \leq B,$$

for $0 \leq t < \min\{T(\varphi), \tau(M)\}$, where B depends on u only through the value of M. We then write $g(u) = hu$, with $h = g(u)/u$ and we apply Lemma 8.1.3, to obtain that, for all $\varepsilon \in (0, \min\{T(\varphi), \tau(M)\}$,

$$\sup_{\varepsilon \leq t < \min\{T(\varphi), \tau(M)\}} \leq K,$$

where K depends on u only through the value of M. It follows that $T(\varphi) > \tau(M)$ and we complete the proof taking $\varepsilon = \tau(M)/2$. □

8.1.2. Uniform estimates

The main result of this section is the following.

Theorem 8.1.6. *Suppose that g satisfies (8.1), and that there exist $M < \infty$ and $\varepsilon > 0$ such that*
$$xg(x) \geq (2+\varepsilon)G(x),$$
for $|x| \geq M$. Let $\varphi \in X$ and let u be the corresponding maximal solution of (5.4) (see Theorem 5.2.1). Then if $T(\varphi) = \infty$, we have $\sup_{t \geq 0} \|u(t)\|_{L^\infty} < \infty$.

Proof. Applying (5.14) and (5.18), we obtain, for all $1 \leq s \leq t < \infty$, the following inequalities:

$$\int_1^t \int_\Omega u_t^2 \, dx \, dt = E(u(1)) - E(u(t)); \tag{8.4}$$

$$\frac{d}{dt} \int_\Omega u(t)^2 \, dx \geq 2(2+\varepsilon) \int_s^t \int_\Omega u_t^2 \, dx \, dt$$
$$+ \varepsilon \int_\Omega |\nabla u(t)|^2 \, dx + k - 2(2+\varepsilon) E(u(s)); \tag{8.5}$$

where $k = 2|\Omega|(\mu - (2+\varepsilon)\nu)$, and μ, ν are defined by Proposition 5.4.4. But if $k - 2(2+\varepsilon)E(u(s)) > 0$, we conclude as in the proof of Proposition 5.4.4 that $T(\varphi) < \infty$ (inequality (5.19) and what follows), which is absurd. We then have $2(2+\varepsilon)E(u(t)) \geq k$ for all $t \geq 1$, and we deduce immediately from (8.4) that

$$\int_1^\infty \int_\Omega u_t^2 \, dx \, dt < \infty. \tag{8.6}$$

Set
$$\Gamma = \left\{ t \geq 1; \int u_t(t)^2 \, dx \leq 1 \right\}.$$

For $t \in \Gamma$, we apply (8.5) with $s = 1$. It follows that

$$\varepsilon \int_\Omega |\nabla u(t)|^2 \, dx \leq 2(2+\varepsilon)E(u(1)) - k + 2 \int_\Omega u u_t \, dx$$
$$\leq 2(2+\varepsilon)E(u(1)) - k + \|u(t)\|_{L^2}. \tag{8.7}$$

From Poincaré's inequality, it follows that

$$\|u(t)\|_{L^2}^2 \leq \frac{\varepsilon}{2} \int_\Omega |\nabla u(t)|^2 \, dx + C(\Omega, \varepsilon).$$

We then deduce from (8.7) that there exists $M < \infty$ such that, for all $t \in \Gamma$, we have $\|u(t)\|_{H^1} \leq M$. By Corollary 8.1.5, we then have

$$\|u(t+s)\|_{L^\infty} \leq K(M), \tag{8.8}$$

for all $t \in \Gamma$ and $s \in [\tau(M)/2, \tau(M)]$. Set $\Sigma_t = \{s \geq t, s \notin \Gamma\}$ for $t \geq 1$. It follows from (8.6) that there exists $\theta < \infty$ such that $|\Sigma_\theta| \leq \tau(M)/4$. We then set $T = \theta + \tau(M)$. Since $u \in C([0,T], X)$, we have in particular

$$\sup_{0 \leq t \leq T} \|u(t)\|_{L^\infty} < \infty.$$

On the other hand, if $t \geq T$ then there exists $s \in [t - \tau(M), t - \frac{1}{2}\tau(M)]$ such that $s \in \Gamma$. Hence we have $t \in [s + \frac{1}{2}\tau(M), s + \tau(M)]$ and so, by (8.8), $\|u(t)\|_{L^\infty} \leq K(M)$. Consequently, $\sup_{t \geq T} \|u(t)\|_{L^\infty} \leq K(M)$. □

Remark 8.1.7. If g does not satisfy (8.1), we do not know whether the conclusions of Theorem 8.1.6 still hold.

The following proposition sharpens the results obtained in §5.3.

Proposition 8.1.8. *Suppose that there exist $\mu < \lambda/2$ (λ given by (2.2)) and $M < \infty$ such that*

$$G(x) \leq \mu x^2,$$

for $|x| \geq M$. Then, for all $\varphi \in X$, the corresponding maximal solution u of (5.4) is global and $\sup_{t \geq 0} \|u(t)\|_{L^\infty} < \infty$.

Proof. By Lemma 5.3.3, we know that u is global. In addition, (8.4) holds, so that

$$\frac{1}{2} \int_\Omega |\nabla u(t)|^2 \, dx \leq E(u(1)) + C + \mu \int_\Omega u(t)^2 \, dx,$$

for $t \geq 1$. Applying (2.2), we then deduce that $\sup_{t \geq 1} \|u(t)\|_{H^1} < \infty$. On the other hand, $g(u)/u$ is bounded in $L^\infty(\Omega)$, and we complete the proof by applying Corollary 8.1.4. □

8.2. The Klein–Gordon equation

In this section, we use the notation from Chapter 6. In particular, Ω is any open subset of $\mathbb{R}^N$, $m > -\lambda$ (where λ is defined by (2.2)), $X = H_0^1(\Omega) \times L^2(\Omega)$, g is a function of $C(\mathbb{R}, \mathbb{R})$ such that $g(0) = 0$, and which satisfies (6.8) with $(N-2)\alpha \leq 2$. G and V are defined by (6.5) and (6.6). E is defined at the beginning of §6.2. We now state the result.

Theorem 8.2.1. *Suppose that $N \geq 3$ and that there exists $\varepsilon > 0$ such that*

$$xg(x) \geq (2+\varepsilon)G(x),$$

for all $x \in \mathbb{R}$. Let $(\varphi, \psi) \in X$ and let u be the corresponding maximal solution of (6.15)–(6.17) (see Theorem 6.2.2). Then, if $T(\varphi, \psi) = \infty$, we have $\sup_{t \geq 0} \|(u(t), u_t(t))\|_X < \infty$.

Proof. We proceed in five steps. Let u be as above and set

$$f(t) = \int_\Omega u(t)^2 \, dx,$$

for $t \geq 0$.

Step 1. Some inequalities. By (6.26), we have

$$f''(t) \geq \varepsilon \left\{ \int_\Omega |\nabla u(t)|^2 \, dx + m \int_\Omega u(t)^2 \, dx \right\} + (4+\varepsilon) \int_\Omega u_t(t)^2 \, dx - 2(2+\varepsilon) E(\varphi, \psi),$$

for all $t \geq 0$. Therefore

$$f''(t) \geq \varepsilon \|u(t)\|_{H^1}^2 + (4+\varepsilon) \int_\Omega u_t(t)^2 \, dx - 2(2+\varepsilon) E(\varphi, \psi), \qquad (8.9)$$

for all $t \geq 0$. In particular, there exist $\eta, \mu > 0$ such that

$$f''(t) \geq \eta f(t) - 2(2+\varepsilon) E(\varphi, \psi), \qquad (8.10)$$
$$f''(t) \geq \mu \|(u(t), u_t(t))\|_{H^1}^2 - 2(2+\varepsilon) E(\varphi, \psi), \qquad (8.11)$$

for all $t \geq 0$. On the other hand, we deduce from the Cauchy–Schwarz inequality that there exists $\delta > 0$ such that

$$\mu \|(u(t), u_t(t))\|_X^2 \geq 2\delta \left| \int_\Omega u(t) u_t(t) \, dx \right| = \delta |f'(t)|.$$

Then it follows from (8.11) that

$$f''(t) \geq \delta |f'(t)| - 2(2+\varepsilon) E(\varphi, \psi), \qquad (8.12)$$

for all $t \geq 0$.

Step 2. We claim that

$$E(\varphi, \psi) \geq 0; \qquad (8.13)$$
$$\frac{d}{dt} (\eta f(t) - 2(2+\varepsilon) E(\varphi, \psi))^+ \leq, \quad \forall t \geq 0, \qquad (8.14)$$
$$\eta f(t) \leq \max \eta f(0), 2(2+\varepsilon) E(\varphi, \psi), \quad \forall t \geq 0. \qquad (8.15)$$

Indeed, if (8.13) does not hold, we know that $T(\varphi, \psi) < \infty$ (Proposition 6.4.1). On the other hand, if (8.14) does not hold, then there exists $t \geq 0$ such that, setting $g(t) = \eta f(t) - 2(2+\varepsilon)E(\varphi, \psi)$, we have

$$g(t) > 0 \quad \text{and} \quad g'(t) > 0. \tag{8.16}$$

Applying (8.10), we also obtain

$$g''(t) \geq \eta g(t), \quad \forall t \geq 0. \tag{8.17}$$

We deduce from (8.16) and (8.17) that $g(t) \to \infty$, as $t \to \infty$. Therefore, by (8.9), there exists $t_0 \geq 0$ such that

$$f''(t) \geq (4+\varepsilon) \int_\Omega u_t(t)^2 \, dx, \quad \forall t \geq t_0.$$

As for Proposition 6.4.1, this implies that $T(\varphi, \psi) < \infty$, which is absurd. We then have (8.14), and (8.15) follows immediately.

Step 3. We claim that

$$\delta |f'(t)| \leq \max\{\delta |f'(0)|, 2(2+\varepsilon)E(\varphi, \psi)\}, \quad \forall t \geq 0. \tag{8.18}$$

Indeed, set $h(t) = \delta f'(t) - 2(2+\varepsilon)E(\varphi, \psi)$, for all $t \geq 0$. We deduce from (8.12) that $h'(t) \geq \delta h(t)$. We then have $h(t) \geq e^{\delta(t-s)} h(s)$, for $t \geq s \geq 0$. If there exists $t \geq 0$ such that $h(t) > 0$, then $h(t) \to \infty$, as $t \to \infty$; and so $f(t) \to \infty$, as $t \to \infty$. This contradicts (8.15), and so

$$\delta f'(t) \leq 2(2+\varepsilon)E(\varphi, \psi), \quad \forall t \geq 0. \tag{8.19}$$

Now set $k(t) = -\delta f'(t) - 2(2+\varepsilon)E(\varphi, \psi)$, for $t \geq 0$. From (8.12), we have $-k'(t) \geq \delta k(t)$. Therefore, $k(t) \leq e^{-\delta t} k(0)$ for $t \geq 0$; and so $k(t) \leq \max\{k(0), 0\}$. Consequently, we have

$$-\delta f'(t) \leq \max\{-\delta f'(0), 2(2+\varepsilon)E(\varphi, \psi)\}, \quad \forall t \geq 0. \tag{8.20}$$

Putting together (8.19) and (8.20), we obtain (8.18).

Step 4. We have

$$\sup_{t \geq 0} \int_t^{t+1} \|(u(s), u_t(s))\|_X^2 \, ds < \infty. \tag{8.21}$$

To verify (8.21), it suffices to integrate (8.11) between t and $t+1$, and next to apply (8.18).

Step 5. Conclusion. Set $w(t) = \|(u(t), u_t(t))\|_X^2$, for $t \geq 0$. We have (see Proposition 6.2.3 and Corollary 6.1.7)

$$w'(t) = 2\int_\Omega g(u(t))u_t(t)\,dx \leq 2\|g(u(t))\|_{L^2}\|u_t(t)\|_{L^2} \\ \leq C\|g(u(t))\|_{L^2} w(t)^{1/2}, \quad (8.22)$$

for all $t \geq 0$. Observe then that $N \geq 3$, and so $N/(N-2) \leq 3$. Consequently,

$$\|g(z)\|_{L^2} \leq C\| |z| + |z|^{\frac{N}{N-2}}\|_{L^2} \leq C\left(\|z\|_{L^2} + \|z\|_{L^{\frac{2N}{N-2}}}^{\frac{N}{N-2}}\right) \\ \leq C\|z\|_{H^1}\left(1 + \|z\|_{H^1}^2\right), \quad (8.23)$$

for all $z \in H_0^1(\Omega)$. It follows from (8.22) and (8.23) that

$$w'(t) \leq Cw(t)(1 + w(t)), \quad \forall t \geq 0. \quad (8.24)$$

We deduce from (8.24) and (8.21) that, for all $0 \leq t \leq s \leq t+1$, we have

$$w(s) \leq Cw(t)\exp\left(\int_t^{t+1} w(\sigma)\,d\sigma\right) \leq Kw(t), \quad (8.25)$$

where K depends neither on t nor on s. In particular, for all $t \geq 1$ and all $\tau \in [0, 1]$, we have

$$w(t) \leq Kw(t - \tau).$$

Integrating this last inequality in τ on $[0, 1]$ and applying (8.21) again, it follows that

$$w(t) \leq K\int_{t-1}^t w(s)\,ds \leq K'.$$

where K' does not depend on t. Since $\sup_{0 \leq t \leq 1} w(t) < \infty$, the proof is complete. □

Remark 8.2.2. If Ω is bounded, we may suppose that g only satisfies the following weaker condition (see Proposition 6.4.4): $xg(x) \geq (2 + \varepsilon)G(x)$, for $|x|$ large.

Remark 8.2.3. We have supposed that $N \geq 3$. Modifying only the end of the proof (Step 5) we can show that the conclusions of Theorem 8.2.1 remain valid if $N = 2$ and $\alpha \leq 4$ (α appears in (6.8)), and if $N = 1$ for any value of $\alpha \geq 0$ (see Cazenave [1] and Sili [1]).

8.3. The non-autonomous heat equation

In this section, we use the notation of Chapter 5. In particular, Ω is a bounded open subset of $\mathbb{R}^N$ with Lipschitz continuous boundary, $X = C_0(\Omega)$, and $(T(t))_{t\geq 0}$ denotes the semigroup associated with the heat equation. g is a locally Lipschitz function $\mathbb{R} \to \mathbb{R}$ such that $g(0) = 0$. On the other hand, we will use the space $H^{-1}(\Omega)$ defined in §2.6.4 and §2.6.5. We also consider $\sigma \geq 1$ such that $\sigma > N/2$. We have in particular $L^s \hookrightarrow H^{-1}$. Given $T > 0$, $\varphi \in X$, and $h : [0, T] \to L^\sigma(\Omega)$, we are going to study the solutions of the following problem:

$$\begin{cases} u \in C([0,T], X) \cap L^1([0,T], H_0^1(\Omega)) \cap W^{1,1}((0,T), H^{-1}(\Omega)); & (8.26) \\ u_t(t) = \Delta u(t) + g(u(t)) + h(t), \quad \text{almost everywhere in } (0, T); & (8.27) \\ u(0) = \varphi. & (8.28) \end{cases}$$

In the following subsection, we gather some preliminary results concerning problem (8.26)–(8.28).

8.3.1. The Cauchy problem for the non-autonomous heat equation

Lemma 8.3.1. Let $T > 0$, $1 \leq p \leq \infty$, and $f \in L^p((0,T), L^\sigma(\Omega))$, and let w be given by

$$w(t) = T(t)\varphi + \int_0^t T(t-s)f(s)\,ds. \qquad (8.29)$$

Then, $w \in C([0,T], L^s) \cap L^p((0,T), H_0^1(\Omega)) \cap L^p((0,T), X)$. If $p = \infty$, we have in addition that $w \in C([0,T], X) \cap C([0,T], H_0^1(\Omega))$.

Proof. Observe first that, by (3.34), the integral appearing in (8.29) does make sense. On the other hand, in view of (3.34) and (3.31), the result is clear if $f \in C([0,T], L^\sigma(\Omega))$. The general case follows by density since, by (3.34), (3.31), and Young's inequality, we have:

$$\|w\|_{L^p((0,T),X)} + \|w\|_{L^p((0,T),H^1)} \leq C\|f\|_{L^p((0,T),L^\sigma)}.$$

Corollary 8.3.2. Let $T > 0$, $1 \leq p \leq \infty$, and $f \in L^p((0,T), L^\sigma(\Omega))$, and $w \in C([0,T], L^\sigma(\Omega))$. Then w is a solution of (8.29) if and only if w is a solution of the following problem:

$$\begin{cases} w \in C([0,T], L^\sigma(\Omega)) \cap L^p([0,T], H_0^1(\Omega)) \cap W^{1,p}((0,T), H^{-1}(\Omega)); & (8.30) \\ w_t(t) = \Delta w(t) + f(t), \quad \text{almost everywhere in } (0, T); & (8.31) \\ w(0) = 0. & (8.32) \end{cases}$$

Proof. Denote by $(S(t))_{t\geq 0}$ the semigroup generated in $H^{-1}(\Omega)$ by the operator C considered in Proposition 2.6.14. It is clear that $(S(t))_{t\geq 0}$ coincides with

$(T(t))_{t\geq 0}$ on $L^\sigma(\Omega)$. In particular, note that $f \in L^p((0,T), H^{-1}(\Omega))$ and then apply Lemma 8.3.1 and Proposition 4.1.9. □

Corollary 8.3.3. *Let $T > 0$, $\varphi \in X$, and $h \in L^\infty((0,T), L^\sigma(\Omega))$, and let $u \in C([0,T], X)$. Then, u is a solution of (8.26)-(8.28) if and only if u is a solution of*

$$u(t) = T(t)\varphi + \int_0^t T(t-s)g(u(s))\,ds + \int_0^t T(t-s)h(s)\,ds. \tag{8.33}$$

In addition, $u \in C((0,T], H_0^1(\Omega)) \cap L^2((0,T), H_0^1(\Omega)) \cap W^{1,2}((0,T), H^{-1}(\Omega))$.

Proof. We apply Lemma 8.3.1 and Corollary 8.3.2 to $f(t) = g(u(t)) + h(t)$ and $w(t) = u(t) - T(t)\varphi$. Thus, we obtain the equivalence between (8.33) and (8.26)–(8.28). On the other hand, it follows easily from (3.18) that $T(\cdot)\varphi \in L^2((0,T), H_0^1(\Omega))$; which implies that $u \in W^{1,2}((0,T), H^{-1}(\Omega))$. By (3.29), we have $T(\cdot)\varphi \in C((0,T], H_0^1(\Omega))$; and so $u \in C((0,T], H_0^1(\Omega))$. □

Now we can state a result of local existence for problem (8.26)–(8.28).

Proposition 8.3.4. *Let $\varphi \in X$, $h \in L_{\text{loc}}^\infty([0,\infty), L^\sigma(\Omega))$. Then, there exists a unique maximal solution $u \in C([0, T(\varphi)), X)$ of (8.33). We have $T(\varphi) > 0$, and if $T(\varphi) < \infty$, then $\|u(t)\|_{L^\infty} \to \infty$ as $t \uparrow T(\varphi)$.*

Proof. Applying Lemma 8.3.1, we easily adapt the proof of Theorem 4.3.4. □

Remark 8.3.5. We see that the condition $g(0) = 0$ is not necessary to solve problem (5.1)–(5.3) or problem (8.26)–(8.28). Indeed, if $g(0) \neq 0$, we can replace g by $g - g(0)$ and h by $h + g(0)1_\Omega$.

8.3.2. A priori estimates

Proposition 8.3.6. *Under the hypotheses of Proposition 8.3.4, and if there exist $M, C \geq 0$ such that*

$$xg(x) \leq Cx^2, \quad \text{for } |x| \geq M, \tag{8.34}$$

then $T(\varphi) = \infty$.

Proof. Let w and z be the maximal solutions of the following problems:

$$w(t) = T(t)\varphi^+ + \int_0^t T(t-s)g(w(s))\,ds + \int_0^t T(t-s)h^+(s)\,ds,$$

$$z(t) = T(t)\varphi^- + \int_0^t T(t-s)(-g(-z(s)))\,ds + \int_0^t T(t-s)h^-(s)\,ds.$$

Applying Lemma 8.1.2, we easily verify that

$$-z(t) \le u(t) \le w(t),$$

for all $t \ge 0$ such that u, w, and z are defined. In addition, applying Lemma 8.1.2 again, we readily obtain $w(t) \ge 0$ and $z(t) \ge 0$. We may restrict ourselves to the case in which $\varphi \ge 0$ and $h \ge 0$, and so $u \ge 0$. We deduce from (8.34) that there exists $C' \ge 0$ such that (see Step 1 of the proof of Proposition 5.3.3)

$$\|g(u(t))^+\|_{L^\infty} \le C' + C\|u(t)\|_{L^\infty},$$

for $t \in [0, T(\varphi))$. Applying (3.34), (3.37), and Lemma 3.5.9, we obtain

$$\|u(t)\|_{L^\infty} \le \|\varphi\|_{L^\infty} + \frac{MC'}{\lambda} + MCe^{-\lambda t}\int_0^t e^{\lambda s}\|u(s)\|_{L^\infty}\,ds$$
$$+ \Big\|\int_0^t \mathcal{T}(t-s)h(s)\,ds\Big\|_{L^\infty}, \qquad (8.35)$$

for $t \in [0, T(\varphi))$. Next, observe that, by (3.34) and (3.37),

$$\|\mathcal{T}(t-s)h(s)\|_{L^\infty} \le Ke^{-\frac{\lambda}{2}(t-s)}(t-s)^{-\frac{N}{2\sigma}}\|h(s)\|_{L^\sigma};$$

and so

$$\Big\|\int_0^t \mathcal{T}(t-s)h(s)\,ds\Big\|_{L^\infty} \le K\|h\|_{L^\infty((0,t),L^\sigma)}\int_0^t e^{-\frac{\lambda}{2}(t-s)}(t-s)^{-\frac{N}{2\sigma}}\,ds$$
$$\le K'\|h\|_{L^\infty((0,t),L^\sigma)}.$$

Therefore, it follows from (8.35) that

$$\|u(t)\|_{L^\infty} \le \|\varphi\|_{L^\infty} + \frac{MC'}{\lambda} + MCe^{-\lambda t}\int_0^t e^{\lambda s}\|u(s)\|_{L^\infty}\,ds$$
$$+ K'\|h\|_{L^\infty((0,t),L^\sigma)}. \qquad (8.36)$$

We conclude by applying Lemma 4.2.1. $\square$

The main result of this section is the following.

Theorem 8.3.7. *Suppose that g satisfies (8.34) with $C < \lambda$, and let $h \in L^\infty(\mathbb{R}_+, L^\sigma(\Omega)) \cap L^\infty(\mathbb{R}_+, L^2(\Omega))$. Then, for all $\varphi \in X$, the maximal solution u of (8.33) is global and satisfies $\sup_{t \ge 0}\|u(t)\|_{L^\infty} < \infty$.*

Proof. We know (Proposition 8.3.6) that $T(\varphi) = \infty$. To establish the bound, as in the proof of Proposition 8.3.6 we may assume that $\varphi \ge 0$ and $h \ge 0$, and so $u \geq 0$. Multiplying the equation by u, integrating by parts, and setting $f(t) = \int_\Omega u(t)^2 \, dx$, it follows that

$$f'(t) \leq -\lambda f(t) + \int_\Omega u(t)g(u(t)) \, dx + \int_\Omega u(t)h(t) \, dx.$$

But $xg(x) \leq C_1 + Cx^2$ and

$$\int_\Omega u(t)h(t) \, dx \leq \|u(t)\|_{L^2}\|h(t)\|_{L^2} \leq \frac{\lambda - C}{2}f(t) + C_2\|h(t)\|_{L^2}^2$$
$$\leq \frac{\lambda - C}{2}f(t) + C_3.$$

Consequently,

$$f'(t) \leq -\frac{\lambda - C}{2}f(t) + C_3.$$

It follows (see Lemma 8.4.6 below) that $\sup_{t \geq 0} \|u(t)\|_{L^2} < \infty$. We conclude as in the proof of Proposition 5.3.6, using (see the proof of Proposition 8.3.6 above)

$$\left\| \int_0^t T(t-s)h(s) \, ds \right\| \leq K'\|h\|_{L^\infty((0,T),L^\sigma)},$$

for $t \in [0, T]$. □

Remark 8.3.8. Applying the estimates of Lemma 8.3.1, we easily show that, for all $\varepsilon > 0$, we have $\sup_{t \geq \varepsilon} \|u(t)\|_{H^1} < \infty$.

8.4. The dissipative non-autonomous Klein–Gordon equation

In this section, we follow the notation of Chapter 6. In particular, Ω is any open subset of $\mathbf{R}^N$, $m > -\lambda$ (where λ is defined by (2.2)), $X = H_0^1(\Omega) \times L^2(\Omega)$, $(S(t))_{t \in \mathbb{R}}$ is the isometry group associated with the Klein–Gordon equation in $Y = L^2(\Omega) \times H^{-1}(\Omega)$, and g is a function of $C(\mathbb{R}, \mathbb{R})$ such that $g(0) = 0$, and which satisfies (6.8) with $(N-2)\alpha \leq 2$. G and V are defined by (6.5) and (6.6). E is defined at the beginning of §6.2. We denote by F the function defined by $F((u,v)) = (0, g(u))$, for $(u,v) \in X$. F is Lipschitz continuous on bounded subsets of X. We also consider $\gamma \geq 0$, and we define the operator $\Gamma \in \mathcal{L}(X)$ given by $\Gamma((u,v)) = (0, \gamma v)$, for $(u,v) \in X$. For $T > 0$, $(\varphi, \psi) \in X$, and $h : [0,T] \to L^2(\Omega)$, we are going to study the solutions of the following problem:

$$\begin{cases} u \in C([0,T], H_0^1(\Omega)) \cap C^1([0,T], L^2(\Omega)) \cap W^{2,1}((0,T), H^{-1}(\Omega)); & (8.37) \\ u_{tt} - \Delta u + mu + \gamma u_t = g(u) + h \text{ in } H^{-1}(\Omega), \quad \text{a.e. in } (0,T); & (8.38) \\ u(0) = \varphi, \quad u_t(0) = \psi. & (8.39) \end{cases}$$

We have a result which is similar to Lemma 6.2.1.

Lemma 8.4.1. *Let $T > 0$, $(\varphi, \psi) \in X$, and $h \in L^1((0,T), L^2(\Omega))$, and let $u \in C([0,T], H_0^1(\Omega)) \cap C^1([0,T], L^2(\Omega))$. Define $H \in L^1((0,T), X)$, $H(t) = (0, h(t))$, for almost every $t \in [0,T]$. Then u is a solution of (8.37)–(8.39) if and only if $U = (u, u_t)$ is a solution of*

$$U(t) = S(t)(\varphi, \psi) + \int_0^t S(t-s)\{F(U(s)) - \Gamma(U(s) + H(s)\}\,ds, \quad (8.40)$$

for all $t \in [0,T]$.

Proof. We apply Proposition 4.1.9 in the space Y. □

Proposition 8.4.2. *Let $h \in L^1_{\mathrm{loc}}(\mathbb{R}_+, L^2(\Omega))$. For all $(\varphi, \psi) \in X$, there exists a unique maximal solution $u \in C([0, T(\varphi, \psi)], H_0^1(\Omega)) \cap C^1([0, T(\varphi, \psi)], L^2(\Omega))$ of (8.37)–(8.39). We have $T(\varphi, \psi) > 0$, and if $T(\varphi, \psi) < \infty$, then*

$$\|(u(t), u_t(t))\|_X \to \infty \quad \text{as } t \uparrow T(\varphi, \psi).$$

In addition,

$$E(u(t), u_t(t)) + \gamma \int_0^t \int_\Omega u_t^2 \, dx\,ds = E(\varphi, \psi) + \int_0^t \int_\Omega h u_t \, dx\, ds, \quad (8.41)$$

for all $t \in [0, T(\varphi, \psi))$.

Proof. We apply the method of Theorem 4.3.4 to solve (8.40), and we apply Lemma 8.4.1 to show (8.41). We note (see the proof of Proposition 4.3.7) that u depends continuously on h and, by density, we need only consider the case in which $h \in C(\mathbb{R}_+, L^2(\Omega))$. In that case, we apply Proposition 6.1.1 and (6.13), and we obtain

$$\frac{d}{dt}E(u(t), u_t(t)) = -\gamma \int_\Omega u_t^2 \, dx + \int_\Omega h u_t \, dx;$$

and hence (8.41). □

Corollary 8.4.3. *Suppose that there exists C such that*

$$G(x) \le Cx^2, \quad \forall x \in \mathbb{R}, \quad (8.42)$$

and let $h \in L^\infty_{\mathrm{loc}}(\mathbb{R}_+, L^2(\Omega))$. Then, for all $(\varphi, \psi) \in X$, we have $T(\varphi, \psi) = \infty$.

Proof. Set $f(t) = \|(u(t), u_t(t))\|_X^2$, for $t \in [0, T(\varphi, \psi))$. It follows from (8.41) and (8.42) that

$$f(t) \le 2E(\varphi, \psi) + 2\int_0^t \|h(s)\|_{L^2} f(s)^{1/2}\, ds + 2C\int_0^t f(s)\, ds,$$

for $t \in [0, T(\varphi, \psi))$; hence the result, applying Lemma 4.2.1. □

Remark 8.4.4. If Ω is bounded, we may assume that (8.42) holds only for $|x|$ large (see Proposition 6.4.4).

The main result of this section is the following.

Theorem 8.4.5. *Suppose that $\gamma > 0$, that g satisfies (8.42) with $2C < \lambda + m$, and that there exist $K \ge 0$ and $c < \lambda + m - KC$ (λ given by (2.2)) such that*

$$xg(x) - KG(x) \le cx^2, \quad \forall x \in \mathbb{R}. \tag{8.43}$$

Let $h \in L^\infty(\mathbb{R}_+, L^2(\Omega))$. Then, for all $(\varphi, \psi) \in X$, we have $T(\varphi, \psi) = \infty$, and the corresponding maximal solution u of (8.37)–(8.39) satisfies $\sup_{t \ge 0} \|(u(t), u_t(t))\|_X < \infty$.

Proof. We know that $T(\varphi, \psi) = \infty$ (Corollary 8.4.3). Take $\varepsilon > 0$ and set

$$f(t) = E(u(t), u_t(t)) + \varepsilon \int_\Omega u u_t\, dx, \quad \forall t \ge 0.$$

It is easy to verify that f is absolutely continuous and that we have

$$f'(t) = \int_\Omega \left\{ -(\gamma - \varepsilon)u_t^2 - \varepsilon|\nabla u|^2 - \varepsilon m u^2 - \varepsilon\gamma u u_t + \varepsilon u g(u) + \varepsilon h u + h u_t \right\} dx,$$

almost everywhere. Let $\delta \in (0, 2\varepsilon)$. We have

$$f' + \delta f = \int_\Omega \Big\{ -\frac{\gamma - \varepsilon - \delta}{2} u_t^2 - \left(\varepsilon - \frac{\delta}{2}\right)|\nabla u|^2$$
$$- m\left(\varepsilon - \frac{\delta}{2}\right) u^2 - \varepsilon(\gamma - \delta)u u_t + \varepsilon u g(u) - \delta G(u) + \varepsilon h u + h u_t \Big\} dx. \tag{8.44}$$

Observe that, by (2.2),

$$-\left(\varepsilon - \frac{\delta}{2}\right)\int_\Omega |\nabla u|^2\, dx - m\left(\varepsilon - \frac{\delta}{2}\right)\int_\Omega u^2\, dx \le -(\lambda + m)\left(\varepsilon - \frac{\delta}{2}\right)\int_\Omega u^2\, dx.$$

On the other hand, if $\delta \leq \varepsilon K$, by applying (8.43) and (8.42), we obtain

$$\varepsilon \int_\Omega ug(u)\,dx - \delta \int_\Omega G(u)\,dx \leq \varepsilon c \int_\Omega u^2\,dx + (\varepsilon K - \delta) \int_\Omega G(u)\,dx$$
$$\leq ((\varepsilon K - \delta)C + \varepsilon c) \int_\Omega u^2\,dx.$$

Thus, it follows from (8.44) that

$$f' + \delta f \leq \int_\Omega \left\{ -\left(\gamma - \varepsilon - \frac{\delta}{2}\right) u_t^2 - \varepsilon\left((\lambda + m)\left(1 - \frac{\delta}{2\varepsilon}\right)\right.\right.$$
$$\left.\left. -\left(K - \frac{\delta}{\varepsilon}\right)C - c\right) u^2 - \varepsilon(\gamma - \delta)uu_t + \varepsilon h u + h u_t \right\} dx. \quad (8.45)$$

Suppose further that $\delta < \gamma$. Applying

$$xy \leq \frac{a}{2}x^2 + \frac{1}{a}y^2, \quad (8.46)$$

with $a = 2\varepsilon(\gamma - \varepsilon)/\gamma$, it follows that

$$-\varepsilon(\gamma - \delta) \int_\Omega uu_t\,dx \leq \frac{\varepsilon^2(\gamma - \delta)^2}{\gamma} \int_\Omega u^2\,dx + \frac{\gamma}{4} \int_\Omega u_t^2\,dx$$
$$\leq \varepsilon^2 \gamma \int_\Omega u^2\,dx + \frac{\gamma}{4} \int_\Omega u_t^2\,dx. \quad (8.47)$$

Applying (8.46) with $a = 1/(2\varepsilon)$, we find that

$$\varepsilon \int_\Omega hu\,dx \leq \frac{1}{4} \int_\Omega h^2\,dx + \varepsilon^2 \int_\Omega u^2\,dx; \quad (8.48)$$

and next, applying (8.46) with $a = 2/\gamma$,

$$\int_\Omega hu_t\,dx \leq \frac{1}{\gamma} \int_\Omega h^2\,dx + \frac{\gamma}{4} \int_\Omega u_t^2\,dx. \quad (8.49)$$

Combining (8.45), (8.47), (8.48), and (8.49), we obtain

$$f' + \delta f \leq \int_\Omega \left\{ -\left(\frac{\gamma}{2} - \varepsilon - \frac{\delta}{2}\right) u_t^2 - \varepsilon\left((\lambda + m)\left(1 - \frac{\delta}{2\varepsilon}\right)\right.\right.$$
$$\left.\left. -\left(K - \frac{\delta}{\varepsilon}\right)C - c - \varepsilon(\gamma + 1)\right) u^2 + \left(\frac{1}{4} + \frac{1}{\gamma}\right) h^2 \right\} dx. \quad (8.50)$$

Note that, for ε sufficiently small, we can take $\delta = \varepsilon^2$. (8.50) then reads

$$f' + \varepsilon^2 f \leq -\left(\frac{\gamma}{2} - \varepsilon - \frac{\varepsilon^2}{2}\right) \int_\Omega u_t^2\,dx - \varepsilon\left((\lambda + m - KC - c)\right.$$
$$\left. -\varepsilon\left(\frac{\lambda + m}{2} - C + \gamma + 1\right)\right) \int_\Omega u^2\,dx + \left(\frac{1}{4} + \frac{1}{\gamma}\right) \int_\Omega h^2\,dx.$$

Recall that we assume that $\lambda + m - KC - c > 0$, and so, if ε is small enough, we have

$$f' + \varepsilon^2 f \leq \left(\frac{1}{4} + \frac{1}{\gamma}\right) \int_\Omega h^2 \, dx. \tag{8.51}$$

In addition, applying (8.42) and (8.46) with $a = 2\varepsilon$, we obtain

$$f(t) \geq \frac{1}{4} \int_\Omega u_t^2 \, dx + \frac{\lambda}{2} \int_\Omega |\nabla u|^2 \, dx + \left(\frac{m}{2} - C - \varepsilon^2\right) \int_\Omega u^2 \, dx.$$

Since $2C < \lambda + m$, applying (2.2), we see that if ε is small enough, then there exists $\delta > 0$ such that

$$f(t) \geq \delta \|((u(t), u_t(t))\|_X^2. \tag{8.52}$$

The result is now a direct consequence of (8.51), (8.52), and of the following lemma. □

Lemma 8.4.6. *Let $T > 0$, $\mu > 0$, and $H \geq 0$. Let $f \in C([0,T])$ be an absolutely continuous function such that*

$$f' + \mu f \leq H,$$

almost everywhere on $(0,T)$. Then, we have

$$f(t) \leq \frac{H}{\mu} + e^{-\mu t} f(0),$$

for all $t \in [0,T]$.

Proof. Set $w(t) = e^{\mu t}(f(t) - H/\mu)$ for $t \in [0,T]$. We have $w'(t) \leq 0$ almost everywhere; and so $w(t) \leq w(0)$ for all $t \in [0,T]$; hence the result. □

Remark 8.4.7. If Ω is bounded, we may suppose that (8.42) and (8.43) hold only for $|x|$ large (see Proposition 6.4.4).

Notes. About §8.1, see Cazenave and Lions [1], Giga [3], and Ni, Sacks, and Tavantzis [1]; and for §8.2, see Cazenave [1] and Sili [1]. Concerning non-autonomous problems (§8.3 and §8.4), see, for example, Haraux [1, 2].

9
The invariance principle and some applications

9.1. Abstract dynamical systems

Throughout this section, (Z, d) is a complete metric space.

Definition 9.1.1. A dynamical system on Z is a family $\{S_t\}_{t \geq 0}$ of mappings on Z such that:

(i) $S_t \in C(Z, Z), \forall t \geq 0$;

(ii) $S_0 = I$;

(iii) $S_{t+s} = S_t \circ S_s, \forall s, t \geq 0$;

(iv) the function $t \mapsto S_t z$ is in $C([0, \infty), Z)$ for all $z \in Z$.

Remark 9.1.2. In what follows, we write $S_t S_s$ instead of $S_t \circ S_s$.

Remark 9.1.3. It is clear that if F is a closed subset of Z such that $S_t F \subset F$ for all $t \geq 0$, then $\{(St)_{|F}\}_{t \geq 0}$ is a dynamical system on (F, d).

Definition 9.1.4. For all $z \in Z$, the continuous curve $t \mapsto S_t z$ is called the trajectory from z.

Definition 9.1.5. Let $z \in Z$. The set

$$\omega(z) = \{y \in Z;\ \exists t_n \to \infty,\ S_{t_n} z \longrightarrow y \text{ as } n \to \infty\},$$

is called the ω-limit set of z.

Proposition 9.1.6. We have $\omega(z) = \bigcap_{s > 0} \overline{\bigcup_{t \geq s} \{S_t z\}}$.

Proof. The proof is straightforward, by Definition 9.1.5. □

Proposition 9.1.7. *For all $z \in Z$ and all $t \geq 0$, we have*

$$\omega(S_t z) = \omega(z), \tag{9.1}$$
$$S_t(\omega(z)) \subset \omega(z). \tag{9.2}$$

In addition, if $\bigcup_{t \geq 0} \{S_t z\}$ is relatively compact in Z, then

$$S_t(\omega(z)) = \omega(z) \neq \emptyset. \tag{9.3}$$

Proof. (9.1) is an immediate consequence of Proposition 9.1.6. Let $y \in \omega(z)$. There exists $t_n \to \infty$ such that $S_{t_n} z \to y$. For all $t \geq 0$, and setting $\tau_n = t_n + t$, we have $S_{\tau_n} z \to S_t y$, and so $S_t y \in \omega(z)$; hence (9.2). Now suppose that $\bigcup_{t \geq 0}\{S_t z\}$ is relatively compact in Z. Then there exists a sequence $t_n \to \infty$ and $y \in Z$ such that $S_{t_n} z \to y$. Therefore $y \in \omega(z)$ and $\omega(z) \neq \emptyset$. It remains to show that $\omega(z) \subset S_t \omega(z)$. To see this, consider $y \in \omega(z)$ and $t_n \to \infty$ such that $S_{t_n} z \to y$. Set $\tau_n = t_n - t$. There exists a subsequence $\tau_{n_k} \to \infty$ such that $S_{\tau_{n_k}} z \to w \in \omega(z)$. Thus,

$$S_t w = S_t \lim S_{\tau_{n_k}} z = \lim S_{t_{n_k}} z = y;$$

hence (9.3). □

Theorem 9.1.8. *Suppose that $\bigcup_{t \geq 0}\{S_t z\}$ is relatively compact in Z. Then:*

(i) $S_t(\omega(z)) = \omega(z) \neq \emptyset$, *for all $t \geq 0$;*

(ii) $\omega(z)$ *is a compact connected subset of Z;*

(iii) $d(S_t z, \omega(z)) \to 0$ *as $t \to \infty$.*

Proof. (i) is a consequence of (9.3). On the other hand, for all $s > 0$, $\bigcup_{t \geq s}\{S_t z\}$ is a relatively compact connected set. By Proposition 9.1.6, $\omega(z)$ is then the decreasing intersection of connected and compacts subsets. Hence we have (ii). To show (iii), assume by contradiction that there exists a sequence $t_n \to \infty$ and $\varepsilon > 0$ such that $d(S_{t_n} z, \omega(z)) \geq \varepsilon$. There exists $y \in Z$ and a subsequence $t_{n_k} \to \infty$ such that $S_{t_{n_k}} z \to y \in \omega(z)$. Therefore $d(S_{t_{n_k}}, \omega(z)) \to 0$ as $k \to \infty$, which is absurd. □

9.2. Liapunov functions and the invariance principle

Definition 9.2.1. *A function $\Phi \in C(Z, R)$ is called a Liapunov function for $\{S_t\}_{t \geq 0}$ if we have*

$$\Phi(S_t z) \leq \Phi(z),$$

for all $z \in Z$ and all $t \geq 0$. •

Remark 9.2.2. If Φ is a Liapunov function for $\{S_t\}_{t\geq 0}$ then, for all $z \in Z$, the function $t \mapsto \Phi(S_t z)$ is non-increasing.

Theorem 9.2.3. (LaSalle Invariance Principle) *Let Φ be a Liapunov function for $\{S_t\}_{t\geq 0}$, and let $z \in Z$ be such that $\bigcup_{t\geq 0}\{S_t z\}$ is relatively compact in Z. Then:*

(i) $\ell = \lim_{t\to\infty} \Phi(S_t z)$ exists;

(ii) $\Phi(y) = \ell$, for all $y \in \omega(z)$.

Proof. $\Phi(S_t z)$ is non-increasing (Remark 9.2.2) and bounded since $\bigcup_{t\geq 0}\{S_t z\}$ is relatively compact. Hence we have (i). If $y \in \omega(z)$, there exists a sequence $t_n \to \infty$ such that $S_{t_n} z \to y$. Therefore, $\Phi(S_{t_n} z) \to \Phi(y)$; hence (ii). □

Definition 9.2.4. An element $z \in Z$ is called an equilibrium point of $\{S_t\}_{t\geq 0}$ if $S_t z = z$ for all $t \geq 0$.

Remark 9.2.5. In practical applications, Theorem 9.2.3 is used mainly to establish that some trajectories of $\{S_t\}_{t\geq 0}$ converge to equilibrium points.

Definition 9.2.6. A Liapunov function Φ for $\{S_t\}_{t\geq 0}$ is said a strict Liapunov function if the following condition is fulfilled. If $z \in Z$ is such that $\Phi(S_t z) = \Phi(z)$ for all $t \geq 0$, then z is an equilibrium point of $\{S_t\}_{t\geq 0}$.

Theorem 9.2.7. *Let Φ be a strict Liapunov function for $\{S_t\}_{t\geq 0}$, and let $z \in Z$ be such that $\bigcup_{t\geq 0}\{S_t z\}$ is relatively compact in Z. Let $\mathcal{E}$ be the set of equilibrium points of $\{S_t\}_{t\geq 0}$. Then:*

(i) $\mathcal{E}$ is a non-empty closed subset of Z;

(ii) $d(S_t z, \mathcal{E}) \to 0$ as $t \to \infty$ (i.e. $\omega(z) \subset E$).

Proof. By continuity of S_t, $\mathcal{E}$ is closed. By Theorem 9.1.8(i), $\omega(z) \neq \emptyset$. Let $y \in \omega(z)$. Applying Theorem 9.1.8(i) again, and then Theorem 9.2.3(ii), we obtain
$$\Phi(S_t y) = \Phi(y), \quad \forall t \geq 0;$$
therefore y is an equilibrium point. From this, we deduce (i) and then (ii) by applying Theorem 9.1.8(iii). □

Remark 9.2.8. Theorem 9.2.7 means that the set of equilibrium points attracts all the trajectories of $\{S_t\}_{t\geq 0}$.

Corollary 9.2.9. *Suppose that the hypotheses of Theorem 9.2.7 are fulfilled. Let $\ell = \lim \Phi(S_t z)$ and $\mathcal{E}_\ell = \{x \in \mathcal{E}, \Phi(x) = \ell\}$. Then $\mathcal{E}_\ell$ is a non-empty closed subset of Z and $d(S_t z, \mathcal{E}_\ell) \to 0$ as $t \to \infty$ (and so $\omega(z) \subset \mathcal{E}_\ell$). If, furthermore, $\mathcal{E}_\ell$ is discrete, then there exists $y \in \mathcal{E}_\ell$ such that $S_t z \to y$ as $t \to \infty$.*

Proof. Since $\mathcal{E}$ is closed and Φ is continuous, $\mathcal{E}_\ell$ is closed. The remaining part of the corollary is a consequence of Theorems 9.2.3, 9.2.7, and 9.1.8 (ii). □

9.3. A dynamical system associated with a semilinear evolution equation

We consider in this section a Banach space X, an m-dissipative operator A with dense domain, and a function $F : X \to X$ that is Lipschitz continuous on bounded subsets. We use the notation of Chapter 4, and in particular we denote by $(T(t))_{t \geq 0}$ the contraction semigroup generated by A. We recall that, for all $x \in X$, there exists a unique maximal solution $u \in C([0, T^*(x)), X)$ of

$$u(t) = T(t)x + \int_0^t T(t-s)F(u(s))\,ds, \quad \forall t \in [0, T^*(x)). \tag{9.4}$$

For $x \in X$ and $t \in [0, T^*(x))$, we set

$$S_t x = u(t).$$

We consider a subset $P \subset X$ such that there exists $M < \infty$ with

$$T(y) = \infty, \quad \forall y \in P; \tag{9.5}$$

$$\|S_t y\| \leq M, \quad \forall y \in P, \forall t \geq 0. \tag{9.6}$$

We set $Z = \overline{\bigcup_{y \in P} \bigcup_{t \geq 0} \{S_t y\}}$, and we denote by d the distance induced on Z by the norm of X.

Lemma 9.3.1. *We have the following properties:*

(i) $T^*(z) = \infty$, $\forall z \in Z$;

(ii) $\|S_t z\| \leq M$, $\forall z \in Z$, $\forall t \geq 0$;

(iii) $S_t z \in Z$, $\forall z \in Z$, $\forall t \geq 0$.

Proof. Let $y \in P$. Then $u(t) = S_t y$ is the solution of (see §4.3)

$$\begin{cases} u \in C([0, \infty), X) \cap C^1([0, \infty), Y); & (9.7) \\ u'(t) = Bu(t) + F(u(t)), \quad \forall t \geq 0; & (9.8) \\ u(0) = y. & (9.9) \end{cases}$$

Therefore, for all $s \geq 0$, $v(t) = u(t+s)$ is the solution of (9.7), (9.8), and $v(0) = u(s)$. Thus, $S_t(S_s y) = S_t(u(s)) = u(t+s)$ for all $s, t \geq 0$. Consequently, we have $T^*(S_s y) = \infty$ for all $y \in P$ and all $s \geq 0$, and $\|S_t S_s y\| \leq M$ for all $y \in P$ and all $s, t \geq 0$. Now take $z \in Z$. There exists a sequence $(t_n)_{n \geq 0} \subset [0, \infty)$ and a sequence $(y_n)_{n \geq 0} \subset P$ such that $S_{t_n} y_n \to z$ as $n \to \infty$. Let $T < T^*(z)$. By Proposition 4.3.7, we have

$$S_t S_{t_n} y_n \longrightarrow S_t z, \quad \text{as } n \to \infty, \tag{9.10}$$

uniformly on $[0, T]$. In particular, we have $\|S_t z\| \leq M$, for $t \in [0, T]$. Since $T < T^*(z)$ is arbitrary, we deduce (i), and next (ii). (iii) is then a consequence of (9.10). □

Theorem 9.3.2. $\{S_t\}_{t \geq 0}$ *is a dynamical system on* (Z, d).

Proof. We have $S_0 = I$. In addition, for all $z \in Z$, if $(z_n)_{n \geq 0} \subset Z$ and $z_n \to z$ as $n \to \infty$ then, by Proposition 4.3.7, we have

$$S_t z_n \longrightarrow S_t z, \quad \text{as } n \to \infty,$$

for all $t \geq 0$. Hence $S_t \in C(Z, Z)$ for all $t \geq 0$. Furthermore, since for all $y \in Z$, $u(t) = S_t y$ is the solution of (9.7)–(9.9), we deduce easily that $S_t S_s = S_{t+s}$ for all $s, t \geq 0$. Finally, we have $S_t z \in C([0, \infty), Z)$ for all $z \in Z$; hence the result. □

9.4. Applications to the non-linear heat equation

We follow the notation of Chapter 5. In particular, Ω is a bounded open subset of $\mathbb{R}^N$, with Lipschitz continuous boundary, $X = C_0(\Omega)$, and $(\mathcal{T}(t))_{t \geq 0}$ denotes the semigroup associated with the heat equation. g is a locally Lipschitz function $\mathbb{R} \to \mathbb{R}$ such that $g(0) = 0$, and we consider G and E to be defined as in Proposition 5.4.4.

Let $\varphi \in X$ be such that $T(\varphi) = \infty$ and let u be the corresponding maximal solution of (5.1)–(5.3) (see Theorem 5.2.1). If we have

$$\sup_{t \geq 0} \|u(t)\|_{L^\infty} < \infty, \tag{9.11}$$

then we may apply the results of §9.3, choosing $Y = \{\varphi\}$ to associate to φ a complete metric (Z, d), where d is the distance induced by the norm in X, $Z = \overline{\bigcup_{t \geq 1} \{u(t)\}}$, and a dynamical system $\{S_t\}_{t \geq 0}$ on (Z, d). On the other hand, we know that there exist sufficient conditions to have (9.11); see, for example, §5.3 and §8.1.

Lemma 9.4.1. *Let φ and u be as above. Then we have the following properties:*

(i) $\bigcup_{t\geq 0} \{u(t)\}$ *is relatively compact in X;*

(ii) *for all $\varepsilon > 0$, we have $\sup_{t\geq\varepsilon} \|u(t)\|_{H^1} < \infty$;*

(iii) *for all $\varepsilon > 0$, $\bigcup_{t\geq\varepsilon} \{u(t)\}$ is relatively compact in $H_0^1(\Omega)$;*

(iv) *E is a strict Liapunov function for $\{S_t\}_{t\geq 0}$.*

Proof. The proof proceeds in three steps.

Step 1. Let $\varepsilon > 0$ and $s \geq 0$. Applying Remark 5.1.2, replacing φ by $u(s)$, we obtain in particular that $\|u(s+\varepsilon)\|_{H^1} \leq C(\varepsilon^{1/2} + \varepsilon^{-1/2})$, where C does not depend on s. Hence we have (ii), since s is arbitrary.

Step 2. To establish (i) and (iii), we need only show that if $t_n \to \infty$, then there exists a subsequence t_{n_k} and $w \in X \cap H_0^1(\Omega)$ such that $u(t_{n_k}) \to w$ in $X \cap H_0^1(\Omega)$ as $k \to \infty$. Set $\tau_n = t_n - 1$, $\varphi_n = u(\tau_n)$ and $u_n(\cdot) = u(\tau_n + \cdot)$. It is clear that $u(t_n) = u_n(1)$. By Step 1, φ_n is bounded in $X \cap H_0^1(\Omega)$, and so there exists $\psi \in L^\infty \cap H_0^1(\Omega)$ and a subsequence (n_k) such that $\varphi_{n_k} \to \psi$ in $L^2(\Omega)$, as $k \to \infty$. Since $\|\varphi_{n_k} - \psi\|_{L^\infty}$ is bounded and $\|\varphi_{n_k} - \psi\|_{L^2} \to 0$, it follows from Hölder's inequality that $\|\varphi_{n_k} - \psi\|_{L^p} \to 0$, for all $1 \leq p < \infty$. In particular, $\varphi_{n_k} \to \psi$ in $L^N(\Omega)$, as $k \to \infty$. From (3.37) and (9.11), we deduce that, for all $k, \ell \in \mathbb{N}$, we have

$$\|u_{n_k} - u_{n_\ell}\|_{L^\infty} \leq t^{-1/2}\|\varphi_{n_k} - \varphi_{n_\ell}\|_{L^N} + \int_0^t \|g(u_{n_k}(s)) - g(u_{n_\ell}(s))\|_{L^\infty}\, ds$$

$$\leq t^{-1/2}\|\varphi_{n_k} - \varphi_{n_\ell}\|_{L^N} + C\int_0^t \|u_{n_k}(s) - u_{n_\ell}(s)\|_{L^\infty}\, ds,$$

for all $t \in (0, 1]$. Consequently (Lemma 8.1.1), $u(t_{n_k}) = u_{n_k}(1)$ is a Cauchy sequence in X. Let w be its limit. Now applying (3.32) and (9.11), we obtain, for all $k, \ell \in \mathbb{N}$, the following inequality:

$$\|u_{n_k} - u_{n_\ell}\|_{H^1} \leq (1 + t^{-1/2})\|\varphi_{n_k} - \varphi_{n_\ell}\|_{L^2}$$

$$+ C\int_0^t (1 + (t-s)^{-1/2})\|u_{n_k}(s) - u_{n_\ell}(s)\|_{L^2}\, ds,$$

for all $t \in (0, 1]$; from this, it follows (Lemma 8.1.1) that $u(t_{n_k}) = u_{n_k}(1)$ is a Cauchy sequence in $H_0^1(\Omega)$; and so that $u(t_{n_k}) \to w$ in $X \cap H_0^1(\Omega)$ as $k \to \infty$. We have shown (i) and (iii).

Step 3. E is a strict Liapunov function on Z. Indeed, E is continuous on $X \cap H_0^1(\Omega)$ and so, by (i) and (iii), E is continuous on (Z, d). Let $z \in Z$ and let $v(t) = S_t z$. It follows from (5.14) that, for all $0 \le s \le t$, we have

$$\int_s^t \int_\Omega v_t^2 \, dx \, d\sigma + E(v(t)) = E(v(s)). \tag{9.12}$$

We then have $E(v(t)) \le E(v(s))$, and consequently E is a Liapunov function. On the other hand, $E(v(t)) = E(z)$ for all $t \ge 0$, and we deduce from (9.12) that

$$\int_0^\infty \int_\Omega v_t^2 \, dx \, d\sigma = 0.$$

Consequently, $v_t = 0$ for almost all $t > 0$, and it follows from this that v is constant in $L^2(\Omega)$, and then is also constant in X. Thus, z is an equilibrium point and E is a strict Liapunov function. This completes the proof. □

Theorem 9.4.2. *Let g be as above. Set $\mathcal{E} = \{u \in D(A); \, \Delta u + g(u) = 0\}$, and $\mathcal{E}_a = \{u \in \mathcal{E}; \, E(u) = a\}$, for $a \in \mathbb{R}$. Let $\varphi \in X$ and let u be the corresponding maximal solution of (5.1)–(5.3) (see Theorem 5.2.1). Suppose that $T(\varphi) = \infty$ and that u satisfies (9.11). Then, we have the following properties:*

(i) $E(u(t))$ converges to a finite limit a, as $t \to \infty$;

(ii) $\mathcal{E}_a \ne \emptyset$;

(iii) $\mathrm{dist}(u(t), \mathcal{E}_a) \to 0$, as $t \to \infty$, where dist denotes the distance in $X \cap H_0^1(\Omega)$.

Proof. We apply Lemma 9.4.1 and Corollary 9.2.9. It suffices to note that the set of equilibrium points of the dynamical system associated with u is included in $\mathcal{E}$. □

Remark 9.4.3. If $N = 1$, we can give a sharper result (see Matano [1]). There exists $w \in \mathcal{E}_a$ such that $u(t) \to w$, as $t \to \infty$. If $N \ge 2$, this remains valid if we suppose that g is analytic (see Simon [1]). In the general case, it remains true for most of the solutions (see Lions [1, 2]) but, except in some special cases (see Louzar [1] and Remarks 9.4.4 and 9.4.5 below), we do not know whether it remains true for any solution, apart from the recent results of Hale and Raugel [1] and Haraux and Poláčik [1].

Remark 9.4.4. If we suppose that $xg(x) \le Cx^2$, with $C < \lambda$ (λ given by (2.2)), then we verify immediately by applying (2.2) that $\mathcal{E} = \{0\}$. In that case, all bounded solutions of (5.1)–(5.3) converge to 0 in $X \cap H_0^1(\Omega)$ as $t \to \infty$.

Remark 9.4.5. If g is strictly concave on $(0, \infty)$, $\mathcal{E} \cap \{u \ge 0\} = \{0, \varphi\}$, where φ is the unique positive solution of $-\Delta \varphi = g(\varphi)$, $\varphi \in H_0^1(\Omega)$. In that case, $\omega(u_0)$ is either 0 or φ, for all $u_0 \ge 0$ (cf. Haraux [5]).

9.5. Application to a dissipative Klein–Gordon equation

In this section, we use the notation of Chapter 6. In particular, Ω is any open subset of $\mathbb{R}^N$, $m > -\lambda$ (where λ is defined by (2.2)), $X = H_0^1(\Omega) \times L^2(\Omega)$, $(S(t))_{t \geq 0}$ is the isometry group associated with the Klein–Gordon equation in $Y = L^2(\Omega) \times H^{-1}(\Omega)$, g is a function of $C(\mathbb{R}, \mathbb{R})$ such that $g(0) = 0$, and which satisfies (6.8) with $(N-2)\alpha < 2$. G and V are defined by (6.5) and (6.6). E is defined at the beginning of §6.2. We denote by F the function defined by $F((u,v)) = (0, g(u))$, for $(u,v) \in X$. F is Lipschitz continuous on bounded subsets of X. We also consider $\gamma \geq 0$, and we define the operator $\Gamma \in \mathcal{L}(X)$ given by $\Gamma((u,v)) = (0, \gamma v)$, for $(u,v) \in X$. For $T > 0$ and $(\varphi, \psi) \in X$, we are going to study the solutions of the following problem:

$$\begin{cases} u \in C([0,T], H_0^1(\Omega)) \cap C^1([0,T], L^2(\Omega)) \cap C^2([0,T], H^{-1}(\Omega)); & (9.13) \\ u_{tt} - \Delta u + mu + \gamma u_t = g(u), \quad \forall t \geq 0; & (9.14) \\ u(0) = \varphi, \quad u_t(0) = \psi. & (9.15) \end{cases}$$

We know (Lemma 8.4.1) that $u \in C([0,T], H_0^1(\Omega)) \cap C^1([0,T], L^2(\Omega))$ is a solution of (9.13)–(9.15) if and only if $U = (u, u_t)$ is a solution of

$$U(t) = S(t)(\varphi, \psi) + \int_0^t S(t-s)\{F(U(s)) - \Gamma(U(s))\}ds, \quad (9.16)$$

for all $t \in [0,T]$. We also know (Proposition 8.4.2) that it is possible to solve locally (9.16) and that the solutions satisfy

$$E(u(t), u_t(t)) + \gamma \int_0^t \int_\Omega u_t^2 = E(\varphi, \psi), \quad (9.17)$$

for all $t \in [0,T]$. In particular, we have $E(u(t), u_t(t)) \leq E(\varphi, \psi)$; and so, if there exists C such that $2C < \lambda + m$ and

$$G(x) \leq Cx^2, \quad \forall x \in \mathbb{R},$$

then we have $T(\varphi, \psi) = \infty$ for all $(\varphi, \psi) \in X$ and

$$\sup_{t \geq 0} \|(u(t), u_t(t))\|_X < \infty,$$

where u is the corresponding solution of (9.13)–(9.15) (see the proof of Proposition 6.3.1, and Remark 6.3.2).

On the other hand, if Ω is bounded, it suffices that g is such that

$$G(x) \leq Cx^2, \quad \text{for } |x| \text{ large.}$$

For the end of this section, it is useful to formulate (9.16) in a different way. To do this, define the operator A_γ on X by

$$\begin{cases} D(A_\gamma) = \{(u,v) \in X;\; \Delta u \in L^2(\Omega), v \in H_0^1(\Omega)\}; \\ A_\gamma(u,v) = (v, \Delta u - mu - \gamma v), \text{ for all } (u,v) \in D(A_\gamma). \end{cases}$$

Lemma 9.5.1. *The operator A_γ is m-dissipative on X, with dense domain. In addition, if we denote by $(T_\gamma(t))_{t \geq 0}$ the contraction semigroup generated by A_γ on X, and if we suppose that $\gamma > 0$, then there exist M and $\sigma > 0$ such that*

$$\|T_\gamma(t)\|_{\mathcal{L}(X)} \leq M e^{-\sigma t}, \tag{9.18}$$

for all $t \geq 0$.

Proof. We show that the operator A_γ is m-dissipative on X, with dense domain, as in Proposition 2.6.9. It then remains to establish (9.18). To do this, we argue as in Theorem 8.4.5. We consider $\varepsilon > 0$, $(\varphi, \psi) \in D(A_\gamma)$. We set $T_\gamma(t)(\varphi, \psi) = (u(t), v(t))$ and

$$f(t) = \frac{1}{2}\int v^2 + \frac{1}{2}\int |\nabla u|^2 + \frac{m}{2}\int u^2 + \varepsilon \int uv.$$

We verify that, for ε small enough, we have $f(t) \geq \delta \|(u(t), v(t))\|_X^2$ and $f' + \varepsilon^2 f \leq 0$. We deduce (9.18), with $\sigma = \varepsilon^2$. □

We verify that $U \in C([0,T], X)$ is a solution of (9.16) if and only if U is a solution of

$$U(t) = T_\gamma(t)(\varphi, \psi) + \int_0^t T_\gamma(t-s) F(U(s))\, ds, \tag{9.19}$$

for all $t \in [0,T]$. Let $(\varphi, \psi) \in X$, and let u be the corresponding maximal solution of (9.13)–(9.15). Suppose that $T(\varphi, \psi) = \infty$ and that $\sup_{t \geq 0} \|(u(t), u_t(t))\|_X < \infty$, and set $Z = \overline{\bigcup_{t \geq 0} \{(u(t), u_t(t))\}}$. The results of §9.3 allow us to associate with u a dynamical system $\{S_t\}_{t \geq 0}$ on (Z, d), where d is the distance induced by the norm in X. We have the following result.

Lemma 9.5.2. *Suppose that Ω is bounded and that $\gamma > 0$. Let $(\varphi, \psi) \in X$ be as above. Then, we have following properties:*

(i) *Z is compact;*

(ii) *E is a strict Liapunov function for $\{S_t\}_{t \geq 0}$.*

Proof. We proceed in four steps. Set $U(t) = (u(t), u_t(t))$ and $H(t) = F(U(t))$, for $t \geq 0$.

Step 1. By (9.19), we have $U(t) = \mathcal{T}_\gamma(t)(\varphi, \psi) + W(t)$, where

$$W(t) = \int_0^t \mathcal{T}_\gamma(s) H(t-s) \, ds.$$

By (Lemma 9.5.1) $\mathcal{T}_\gamma(t)(\varphi, \psi) \to 0$ in X, as $t \to \infty$, and so there exists a compact subset K_1 of X such that $\bigcup_{t \geq 0} \{\mathcal{T}_\gamma(t)(\varphi, \psi)\} \subset K_1$. Then we need only verify that there exists a compact subset K_2 of X such that $\bigcup_{t \geq 0} \{W(t)\} \subset K_2$.

Step 2. Since Ω is bounded, we see by applying Theorem 1.3.2 and Remark 1.3.3, as well as the estimates of §6.1.2 (see in particular the proof of Proposition 6.1.5), that the range by the mapping $u \mapsto g(u)$ of a bounded subset of $H_0^1(\Omega)$ is a relatively compact subset of $L^2(\Omega)$. Since u is bounded in $H_0^1(\Omega)$, there exists a compact K of X such that $\bigcup_{t \geq 0} \{H(t)\} \subset K$.

Step 3. Let $\varepsilon > 0$, and let T be such that (see Lemma 9.5.1)

$$\|H\|_{L^\infty(0,\infty;X)} \int_T^\infty \|\mathcal{T}_\gamma\|_{\mathcal{L}(X)} < \varepsilon.$$

For $t \geq T$, we then have

$$\left\| W(t) - \int_0^T \mathcal{T}_\gamma(s) H(t-s) \, ds \right\|_X < \varepsilon.$$

Consequently,

$$\bigcup_{t \geq T} \{W(t)\} \subset K' + B(0, \varepsilon), \tag{9.20}$$

where we have set

$$K' = \bigcup_{t \geq T} \left\{ \int_0^T \mathcal{T}_\gamma(s) H(t-s) \, ds \right\}.$$

Observe that the mapping $(s, x) \mapsto \mathcal{T}_\gamma(s) x$ is continuous from $[0, \infty) \times X$ to X. Consequently, $U = \bigcup_{0 \leq t \leq T} \{\mathcal{T}_\gamma(t) K\}$ is compact in X. Therefore, $F = T \cdot \text{conv}(U)$ is relatively compact in X. Since $K' \subset F$, K' is relatively compact in X. By (9.20), we can cover $\bigcup_{t \geq T} \{W(t)\}$ by a finite union of balls of radius ε. On the other hand, $W \in C([0, \infty), X)$; hence $\bigcup_{0 \leq t \leq T} \{W(t)\}$ is compact and it can also be covered by a finite union of balls of radius ε. Finally, we can cover $\bigcup_{t\geq 0}\{W(t)\}$ by a finite union of balls of radius ε. Since ε is arbitrary, $\bigcup_{t\geq 0}\{W(t)\}$ is relatively compact and, by applying Step 1, we obtain (i).

Step 4. E is continuous on X, and thus also on (Z,d). (9.17) shows that E is a Liapunov function. Finally, if E is constant on a trajectory (v,v_t) of $\{S_t\}_{t\geq 0}$, we deduce from (9.17) that $v_t = 0$ for all $t \geq 0$; therefore v does not depend on t; $z = (v,0)$ is then an equilibrium point, and E is strict Liapunov function. This completes the proof. □

Theorem 9.5.3. *Suppose that Ω is bounded and that $\gamma > 0$. Set $\mathcal{E} = \{u \in H_0^1(\Omega); \Delta u - mu + g(u) = 0\}$, and $\mathcal{E}_a = \{u \in \mathcal{E}; E(u,0) = a\}$, for $a \in \mathbb{R}$. Let $(\varphi,\psi) \in X$, and let u be the corresponding maximal solution of (9.13)–(9.15). Suppose that $T(\varphi,\psi) = \infty$ and that $\sup_{t\geq 0}\|(u(t),u_t(t))\|_X < \infty$. Then, we have the following properties:*

(i) $E(u(t),u_t(t))$ *converges to a finite limit β, as $t \to \infty$;*

(ii) $\mathcal{E}_\beta \neq \emptyset$;

(iii) $\|u_t(t)\|_{L^2} \to 0$, *as $t \to \infty$;*

(iv) dist(u(t),$\mathcal{E}_\beta$) $\to 0$, *as $t \to \infty$, where dist denotes the distance in $H_0^1(\Omega)$.*

Proof. We apply Lemma 9.4.1 and Corollary 9.2.9. It suffices to observe that the set of equilibrium points of the dynamical system associated with u is included in $\mathcal{E}$. □

Remark 9.5.4. If we suppose that $xg(x) \leq Cx^2$, with $C < \lambda + m$ (λ given by (2.2)), then we verify immediately by applying (2.2) that $\mathcal{E} = \{0\}$. In that case, all bounded solutions of (9.13)–(9.15) converge to 0 in X as $t \to \infty$. If g does not satisfy this condition, sufficient conditions to ensure that $\omega(\varphi,\psi)$ is reduced to one point can be found in the literature (cf. Haraux [4], Hale and Raugel [1]).

Notes. See Ball [3], Dafermos [1–3], Dafermos and Slemrod [1], Hale [1], Haraux [1, 2], Henry [1], LaSalle [1], and Sell [1]. The ω-limit sets also appear in the theory of maximal attractors. Consult Babin and Vishik [1–3], Ghidaglia and Temam [1, 2], Hale [2], Haraux [2], and Ladyzhenskaya [2]. The invariance principle is also very useful in the study of the behaviour at infinity of positive solutions of reaction–diffusion systems. See, for example, Masuda [1], Haraux and Kirane [1], and Haraux and Youkana [1]. There has recently been substantial progress on asymptotic behaviour of gradient-like systems as a consequence of the work of Hale and Raugel [1] (cf., e.g., Haraux and Poláčik [1] where the condition of Hale and Raugel is used in an essential way). On the other hand, *negative* results are beginning to appear in the literature (see Poláčik and Rybakowsky [1]) when the non-linearity depends on x.

10
Stability of stationary solutions

In this chapter, we describe an extension of the Liapunov linearization method to establish the (local or global) asymptotic stability of equilibria. The perturbation argument developed here is applicable to various semilinear evolution problems on infinite-dimensional Banach spaces. We also discuss the connection between stability and positivity in the case of the heat equation.

10.1. Definitions and simple examples

Let (X, d) be a complete metric space and $\{S(t)\}_{t\geq 0}$ a dynamical system on X.

Definition 10.1.1. A trajectory $v(t) = S(t)a$ of the dynamical system $\{S(t)\}_{t\geq 0}$ is called (positively) stable in the sense of Liapunov if

$$\forall \varepsilon > 0, \ \exists \delta > 0 \text{ such that}$$
$$x \in X \text{ and } d(x, a) \leq \delta \Rightarrow \forall t \geq 0, \ d(S(t)x, v(t)) \leq \varepsilon. \tag{10.1}$$

Definition 10.1.2. A trajectory $v(t) = S(t)a$ of the dynamical system $\{S(t)\}_{t\geq 0}$ is called (positively) asymptotically stable in the sense of Liapunov if it is stable in the sense of Liapunov and

$$\exists \delta_1 \geq 0 \text{ such that } x \in X, \ d(x, a) \leq \delta_1 \Rightarrow \lim_{t\to\infty} d(S(t)x, v(t)) = 0. \tag{10.2}$$

In particular, an equilibrium a of $\{S(t)\}_{t\geq 0}$ is called stable (resp. asymptotically stable) in the sense of Liapunov if the constant trajectory $v(t) \equiv a$ satisfies (10.1) (resp. (10.1) and (10.2)).

In the easiest cases for $X = \mathbb{R}^N$, stability of the equilibrium a of an equation

$$u' = f(u(t)), \quad t \geq 0, \tag{10.3}$$

with $f \in C^1(X, X)$ can be seen from the linearized equation

$$z' = (Df)(a)(z(t)), \quad t \geq 0. \tag{10.4}$$

More precisely, the exponential asymptotic stability of a for (10.3) is related to exponential asymptotic stability of 0 for the linearized equation (10.4). We recall here the Liapunov stability theorem (for a proof, cf., e.g., Haraux [5]).

Theorem 10.1.3. (Liapunov) Let X be a finite-dimensional normed space, and $f \in C^1(X, X)$ a vector field on X. Let $a \in X$ be such that $f(a) = 0$ and assume that

$$\text{all eigenvalues } s_j, \ 1 \leq j \leq k \text{ of } Df(a) \text{ have negative real parts.} \quad (10.5)$$

Then a is an asymptotically Liapunov stable equilibrium solution of equation (10.3) in the following sense: for each $\delta < \nu = \min_{1 \leq j \leq k}\{-\text{Re}(s_j)\}$, there exists $\rho = \rho(\delta) > 0$ and $M(\delta) \geq 1$ such that, if $\|x - a\| \leq \rho(\delta)$, the solution u of (10.3) such that $u(0) = x$ is global with

$$\forall t \geq 0, \ \|u(t) - a\| \leq M(\delta)\|x - a\|e^{-\delta t}.$$

In the opposite direction, we have the following result (cf. Haraux [5]).

Proposition 10.1.4. Let X be a finite-dimensional normed space, and $f \in C^1(X, X)$ a vector field on X. Let $a \in X$ be such that $f(a) = 0$ and assume that

$$\text{all eigenvalues } s_j, \ 1 \leq j \leq k \text{ of } Df(a) \text{ have positive real parts.} \quad (10.6)$$

Then a is a completely unstable equilibrium solution of (10.3) in the following sense: there is a neighbourhood ω of a in X such that, for each $b \in X$, $b \neq a$, the unique solution of (10.3) with initial condition b leaves ω for ever if $t \geq T$ large enough.

Remark. Stability of an equilibrium cannot always be seen on the linearization. As a simple example, we may consider $u' = f(u) = \pm u^3$, in which cases $Df(0) = 0$. When we choose the $(-)$ sign, 0 is asymptotically (but not exponentially) stable, while when we choose the $(+)$ sign it becomes completely unstable. In this last case, 0 is the only global solution.

To illustrate the general ideas of this section, we give two simple examples.

Example 10.1.5. Let $f \in C^1(\mathbb{R})$ and consider the first order *scalar* ODE

$$u'(t) = f(u(t)).$$

It is known (cf., e.g., Haraux [5]) that each bounded global solution $u(t)$ of this equation on $\mathbb{R}^+$ tends to a limit c with $f(c) = 0$. The stability of such an equilibrium c is delicate only when $f'(c) = 0$. Indeed,

- If $f'(c) < 0$, c is exponentially stable.

156 *Stability of stationary solutions*

- If $f'(c) > 0$, c is completely unstable in the sense of Proposition 10.1.4.

As an illustration, the simple first order ODE

$$u' + u^3 - u = 0$$

has exactly three equilibria $\{-1, 0, 1\}$. The equilibria 1 and (-1) are exponentially stable and they attract, respectively, the positive solutions and the negative solutions of the equation. On the other hand, the equilibrium 0 is completely unstable in a very strong sense: it attracts no solution except itself.

Example 10.1.6 Let $f \in C^1(\mathbb{R})$, $\alpha > 0$ and consider the second order ODE

$$u''(t) + \alpha u'(t) = f(u(t)).$$

It is known (cf., e.g., Haraux [5], Hale and Raugel [1]) that each bounded global solution $u(t)$ of this equation on $\mathbb{R}^+$ tends to a limit c such that $f(c) = 0$ (and $u'(t)$ tends to 0). The stability here is defined in the sense of the phase space $\mathbb{R} \times \mathbb{R}$ for the corresponding first order system in (u, u'). The situation is more complicated than in the previous example:

- If $f'(c) < 0$, then $(c, 0)$ is exponentially stable in the phase space $\mathbb{R} \times \mathbb{R}$.

- If $f'(c) > 0$, then $(c, 0)$ is unstable in the phase space $\mathbb{R} \times \mathbb{R}$ but attracts some other trajectories than the equilibrium itself. We have here a typical example of a *hyperbolic point*.

As an illustration, the simple second order ODE

$$u'' + u' + u^3 - u = 0$$

has exactly three equilibria $\{(-1, 0); (0, 0); (1, 0)\}$. The equilibria $(1, 0)$ and $(-1, 0)$ are exponentially stable in the phase space $\mathbb{R} \times \mathbb{R}$. On the other hand, the equilibrium $(0,0)$ is a *hyperbolic point*.

10.2. A simple general result

Let X be a real Banach space, let $\mathcal{T}(t) = e^{ct}S(t)$ with $c \in \mathbb{R}$, and let $(S(t))_{t \geq 0}$ be a contraction semigroup on X (it is easy to check that the family of operators $(\mathcal{T}(t))_{t \geq 0}$ has the semigroup property, cf. Definition 3.4.1), and $F : X \to X$ locally Lipschitz continuous on bounded subsets. For any $x \in X$, we consider the unique maximal solution $u \in C([0, \tau(x)), X)$ of the equation

$$u(t) = \mathcal{T}(t)x + \int_0^t \mathcal{T}(t-s)F(u(s))\mathrm{d}s, \quad \forall t \in [0, \tau(x)). \tag{10.7}$$

By a stationary solution of (10.7) we mean a constant vector $a \in X$ such that

$$a = T(t)a + \int_0^t T(t-s)F(a)\,ds, \quad \forall t \geq 0. \tag{10.8}$$

The following result is an easy consequence of the general theory of strongly continuous linear semigroups, and can easily be verified. Let $L = cI + A$, where A is the generator of $S(t)$ (L can be considered the generator of $T(t)$ in the sense of Definition 3.4.2). Then we have the following.

Lemma 10.2.1. *A vector $a \in X$ is a stationary solution of (10.7) if and only if we have*

$$a \in D(L) \quad \text{and} \quad La + F(a) = 0. \tag{10.9}$$

We are now in a position to state the main result of this section.

Theorem 10.2.2. *Assume that, for some constants $\delta > 0$, $M \geq 1$, we have*

$$\forall t \geq 0, \quad \|T(t)\| \leq M e^{-\delta t}. \tag{10.10}$$

Let $a \in X$ be a stationary solution of (10.7) such that

$$\exists R_0 > 0,\ \exists \nu > 0: \quad \|F(u) - F(a)\| \leq \nu \|u - a\| \text{ for } \|u - a\| \leq R_0, \tag{10.11}$$

with

$$\nu < \delta/M. \tag{10.12}$$

Then, for all $x \in X$ such that

$$\|x - a\| \leq R_1 = R_0/M, \tag{10.13}$$

the solution u of (10.7) is global and satisfies

$$\forall t \geq 0, \quad \|u(t) - a\| \leq M \|x - a\| e^{-\gamma t}, \tag{10.14}$$

with $\gamma = \delta - \nu M > 0$.

Proof. On replacing u by $u - a$ and F by $F - F(a)$, we may assume that $a = 0$ and $F(a) = 0$ with $\|F(u)\| \leq \nu \|u\|$ whenever $\|u\| \leq R_0$. In particular, setting

$$T = \sup\{t \geq 0,\ \|u(t)\| \leq R_0\} \leq \infty,$$

we find that

$$\forall t \in [0, T), \quad \|u(t)\| \leq M\|x\|e^{-\delta t} + \nu M \int_0^t e^{-\delta(t-s)} \|u(s)\|\,ds.$$

Letting $\varphi(t) = e^{\delta t}\|u(t)\|$, we obtain

$$\varphi(t) \leq C_1 + C_2 \int_0^t \varphi(s)\,ds, \quad \text{for all } t \in (0,T)$$

with $C_1 = M\|x\|$ and $C_2 = \nu M$.

By applying Gronwall's lemma, we deduce that

$$\forall t \in [0,T), \quad e^{\delta t}\|u(t)\| \leq M\|x\|e^{\nu M t}. \tag{10.15}$$

Since $\delta > \nu M$, we conclude that if $M\|x\| \leq R_0$, then $T = +\infty$ and (10.15) holds true on $[0,\infty)$. This completes the proof of (10.14). □

Remark 10.2.3. It is not sufficient for our purposes to state Theorem 10.2.2 with $c \leq 0$ (in which case, $\mathcal{T}(t)$ itself is a contraction semigroup). Indeed, in the examples given below in §10.3, the generator of the linearized equation will not be dissipative in general, especially when working in $C_0(\Omega)$.

10.3. Exponentially stable systems governed by PDE

In this paragraph, we show how the stability theorem 10.2.2 can be applied to partial differential equations.

(a) We first consider the semilinear heat equation

$$\begin{cases} u_t - \Delta u + f(u) = 0 & \text{in } \mathbb{R}^+ \times \Omega, \\ u = 0 & \text{on } \mathbb{R}^+ \times \partial\Omega \end{cases} \tag{10.16}$$

where Ω is a bounded domain in $\mathbb{R}^N$ and f is a function of class $C^1 : \mathbb{R} \to \mathbb{R}$ such that

$$f(0) = 0 \quad \text{and} \quad f'(0) > -\lambda_1, \tag{10.17}$$

where $\lambda_1 = \lambda_1(\Omega)$ is the smallest eigenvalue of $(-\Delta)$ in $H_0^1(\Omega)$. We have the following simple result.

Proposition 10.3.1. *Under the above hypotheses, the stationary solution $u \equiv 0$ of (10.16) is exponentially stable in $X = C_0(\Omega)$ in the following sense: for each $\gamma \in (0, \lambda_1 + f'(0))$, there exists $R = R(\gamma)$ such that for all $x \in X$ with $\|x\| \leq R$, the solution u of (10.16) such that $u(0) = x$ exists and is global, and satisfies*

$$\forall t \geq 0, \quad \|u(t)\| \leq M\|x\|e^{-\gamma t}, \tag{10.18}$$

with M independent of γ and x.

Proof. We have shown in Corollary 3.5.10 that the contraction semigroup $T_0(t)$ generated in $C_0(\Omega)$ by the equation

$$\begin{cases} u_t - \Delta u = 0 & \text{in } \mathbb{R}^+ \times \Omega, \\ u = 0 & \text{on } \mathbb{R}^+ \times \partial\Omega \end{cases}$$

satisfies (10.10) with $\delta = \lambda_1$ and some $M > 1$. It is therefore sufficient to apply Theorem 10.2.2 with $T(t) = e^{-f'(0)t}T_0(t)$, since for $f \in C^1(\mathbb{R})$, $F(u) = f(u) - f'(0)u$ satisfies (10.11) with $a = 0$ and ν arbitrarily small. □

(b) Another situation: this time we assume some conditions which are in a sense opposite to (10.17):

$$f \text{ is strictly convex on } [0, \infty) \text{ and } f(0) = 0, \ f'_d(0) < -\lambda_1(\Omega) \qquad (10.19)$$

where $\lambda_1(\Omega)$ is the smallest eigenvalue of $(-\Delta)$ in $H_0^1(\Omega)$. Here the solution 0 is unstable and we have the following.

Theorem 10.3.2. *(i) There exists one and only one positive solution φ of the problem:*

$$\varphi \in X \cap H_0^1(\Omega), \quad -\Delta\varphi + f(\varphi) = 0. \qquad (10.20)$$

(ii) For each $u_0 \in X$, $u_0 \geq 0$ and not identically 0, the solution u of (10.16) such that $u(0) = u_0$ tends to φ as $t \to \infty$. Moreover, we have

$$\forall t \geq 0, \quad \|u(t,.) - \varphi(t,.)\|_{L^\infty} \leq C(u_0)\exp(-\gamma t), \qquad (10.21)$$

where $\gamma > 0$ is independent of u_0.

Proof. The proof is divided into several steps.

Step 1. We already know by Theorem 9.4.2 and the positivity preserving property that the solution $u(t,.)$ asymptotes towards the set of non-negative solutions of (10.20) as $t \to \infty$. We now show that if $u_0 \neq 0$, $u(t,.)$ cannot tend to 0 as $t \to \infty$. Indeed, assuming that $\lim_{t\to\infty} \|u(t,.)\|_{L^\infty} = 0$, then, for each $\varepsilon > 0$, there is $T(\varepsilon)$ such that

$$\forall t \geq T(\varepsilon), \quad f(u(t,x)) \leq \{f'_d(0) + \varepsilon\}u(t,x) \quad \text{on } \Omega.$$

Choosing $\varepsilon > 0$ so small that $-f'_d(0) - \varepsilon - \lambda_1(\Omega) > 0$, multiplying the equation by a positive eigenfunction φ_1 corresponding to the first eigenvalue $\lambda_1(\Omega)$ of $(-\Delta)$ in $H_0^1(\Omega)$, and then integrating over Ω, we find,

$$\frac{d}{dt}\int_\Omega u(t,x)\varphi_1(x)\,dx \geq 0 \quad \text{for all } t \geq T(\varepsilon).$$

160   *Stability of stationary solutions*

Since the function: $t \to \int_\Omega u(t,x)\varphi_1(x)\,dx$ is non-decreasing on $[T(\varepsilon),\infty)$ and tends to 0 as $t \to \infty$, it must vanish identically on $[T(\varepsilon),\infty)$. Because φ_1 is positive on Ω, this implies that $u(t,.) = 0$ for all $t \geq T(\varepsilon)$. Then a classical connectedness argument shows that $u_0 = 0$. Therefore if $u_0 \neq 0$, the ω-limit set of u_0 under $S(t)$ contains at least a non-negative solution $\varphi \neq 0$ of (10.20).

Step 2.   By the strong maximum principle, we must have $f(\varphi) \leq 0$ and then $\varphi > 0$ in Ω. We now prove the following.

Lemma 10.3.3.   *Let f be as above and $\varphi > 0$ in Ω be a solution of the equation*

$$\varphi \in C(\Omega) \cap H_0^1(\Omega), \quad -\Delta\varphi + f(\varphi) = 0.$$

On the other hand, let $\psi \geq 0$ satisfy

$$\psi \in C(\Omega) \cap H_0^1(\Omega), \quad -\Delta\psi + f(\psi) \geq 0.$$

Then either $\psi = 0$ or $\psi \geq \varphi$.

Proof.   We first establish

$$\forall w \in H_0^1(\Omega), \ \int_\Omega \{|\nabla w|^2 + k(\varphi)w^2\}dx \geq 0, \quad \text{with } k(\varphi) := f(\varphi)/\varphi. \tag{10.22}$$

In fact, denoting by $\mathcal{D}(\Omega)$ the set of real-valued C^∞ functions with compact support in Ω, we have the following sequence of identities

$$\forall w \in \mathcal{D}(\Omega), \ \int_\Omega \{|\nabla w|^2 + k(\varphi)w^2\}dx = \int_\Omega \{|\nabla w|^2 + (\Delta\varphi/\varphi)w^2\}dx$$

$$= \int_\Omega \{|\nabla w|^2 - \nabla\varphi \cdot \nabla(w^2/\varphi)\}dx.$$

Since $\nabla(w^2/\varphi) = 2(w\nabla w/\varphi) - (w^2/\varphi^2)\nabla\varphi$, we obtain the formula

$$\forall w \in \mathcal{D}(\Omega), \ \int_\Omega \{|\nabla w|^2 + [f(\varphi)/\varphi]w^2\}dx = \int_\Omega |\nabla w - (w/\varphi)\nabla\varphi|^2 \, dx. \tag{10.23}$$

This establishes (10.22) when $w \in \mathcal{D}(\Omega)$. Then, by passing to the limit in (10.22) in the sense of $H_0^1(\Omega)$ along a sequence of functions $w_n \in \mathcal{D}(\Omega)$ tending to w, we find that

$$\forall w \in H_0^1(\Omega), \ \int_\Omega \{|\nabla w|^2 + [f(\varphi)/\varphi]w^2\}dx \geq 0.$$

We may, in particular, use (10.22) with $w = (\varphi - \psi)^+ \in C(\Omega) \cap H_0^1(\Omega)$. On the other hand, we have, by the properties of φ and ψ:

$$-\Delta(\varphi - \psi) + f(\varphi) - f(\psi) \leq 0.$$

Multiplying by $w = (\varphi - \psi)^+$ and integrating over Ω, we find that

$$\int_\Omega \{|\nabla w|^2 + [(f(\varphi) - f(\psi))/(\varphi - \psi)]w^2\}dx \leq 0. \tag{10.24}$$

By combining (10.22) and (10.24), we finally obtain

$$\int_\Omega \{[(f(\varphi) - f(\psi))/(\varphi - \psi)] - [f(\varphi)/\varphi]\}w^2\,dx \leq 0. \tag{10.25}$$

As a consequence of strict convexity of f, (10.25) now implies that

$$w\psi = \psi(\varphi - \psi)^+ = 0, \quad \text{everywhere in } \Omega. \tag{10.26}$$

Since $\varphi > 0$ in Ω and $\psi \in C(\Omega)$, the conclusion follows at once from (10.26). □

Lemma 10.3.3 implies in particular the uniqueness part of (i). By combining this with Step 1 we conclude that $\omega(u_0) = \{\varphi\}$, which means that $u(t,.)$ tends to φ as $t \to \infty$.

Step 3. We now establish (10.21). We begin with the identity

$$\frac{d}{dt}\int_\Omega |u - \varphi|^2\,dx = -2\int_\Omega \{|\nabla(u - \varphi)|^2 + \frac{f(u) - f(\varphi)}{u - \varphi}|u - \varphi|^2\}dx. \tag{10.27}$$

From the convergence of $u(t,.)$ to φ in X, it follows in particular that, fixing some non-empty open set ω contained in a compact subset of Ω, we have for $t \geq T$ (depending on the solution u),

$$\forall t \geq T, \quad u(t,x) \geq (1/2)\varphi(x) \quad \text{in } \omega. \tag{10.28}$$

Now, from (10.27) and (10.28), we easily deduce the inequality

$$\frac{d}{dt}\int_\Omega |u - \varphi|^2\,dx \leq -2\int_\Omega \{|\nabla(u - \varphi)|^2 + c(x)|u - \varphi|^2\}dx, \tag{10.29}$$

with

$$c(x) = \begin{cases} 2\dfrac{f(\varphi) - f(\varphi/2)}{\varphi} & \text{if } x \in \omega, \\ 0 & \text{if } x \notin \omega. \end{cases}$$

The result will now become a consequence of the following lemma.

Lemma 10.3.4. *There exists $\delta > 0$ so that, $c(x)$ being given as above, we have the inequality*

$$\forall w \in H_0^1(\Omega), \quad \int_\Omega \{|\nabla w|^2 + c(x)w^2\}dx \geq \delta \int_\Omega w^2\,dx. \tag{10.30}$$

Proof. We introduce

$$\delta = \inf\left\{\int_\Omega \{|\nabla w|^2 + c(x)w^2\}\mathrm{d}x, \quad w \in H_0^1(\Omega), \quad \int_\Omega w^2\,\mathrm{d}x = 1\right\}. \quad (10.31)$$

Since $c \in L^\infty(\Omega)$, a standard argument shows that the infimum in (10.31) is achieved for $w = \zeta \geq 0$, where $\zeta \in H_0^1(\Omega)$ satisfies $\int_\Omega \zeta^2\,\mathrm{d}x = 1$ and is a solution of the elliptic problem

$$\zeta \in C(\Omega) \cap H_0^1(\Omega), \quad -\Delta\zeta + c(x)\zeta = \delta\zeta. \quad (10.32)$$

Multiplying by φ and integrating over Ω, we immediately obtain

$$\delta \int_\Omega \zeta\varphi\,\mathrm{d}x = \int_\Omega (-\Delta\zeta + c(x)\zeta)\varphi\,\mathrm{d}x = \int_\Omega (-\Delta\varphi\,\zeta + c(x)\zeta\varphi)\mathrm{d}x$$
$$= \int_\Omega [c(x) - k(\varphi)(x)]\zeta(x)\varphi(x)\,\mathrm{d}x,$$

where $k(\varphi) = f(\varphi)/\varphi$. By the strict convexity of f, it now follows that $c(x) - k(\varphi)(x) \geq 0$ in Ω and $c(x) - k(\varphi)(x) > 0$ in ω. In addition, we have $\zeta > 0$ everywhere in Ω by (10.32) and the strong maximum principle: in particular, we find $\delta > 0$. The result (10.30) follows at once by homogeneity. □

Proof of Theorem 10.3.2 continued. We deduce from (10.29) and (10.30) the simple inequality

$$\frac{\mathrm{d}}{\mathrm{d}t}\left(\|u(t) - \varphi\|_2^2\right) \leq -2\delta\|u(t,.) - \varphi\|_2^2. \quad (10.33)$$

From (10.33), we first deduce that

$$\forall t \geq T, \quad \|u(t,.) - \varphi\|_2 \leq \|u(T,.) - \varphi\|_2.$$

In fact, (10.27) and the convexity of f also implies that $\|u(t,.) - \varphi\|_2$ is non-increasing; hence

$$\forall t \geq T, \quad \|u(t,.) - \varphi\|_2 \leq \exp(\delta T)\|u_0 - \varphi\|_2 \exp(-\delta t)$$
$$\leq K\|u_0 - \varphi\|_\infty \exp(-\delta t), \quad (10.34)$$

for some $K > 0$. Then, since u and z remain bounded in C^1, from (10.34), we deduce that

$$\forall t \geq 0, \quad \|u(t,.) - \varphi\|_\infty \leq C(u_0)\exp(-\gamma t), \quad (10.35)$$

by replacing δ by a slightly smaller positive constant, denoted by γ. Hence Theorem 10.3.2 is completely proven. □

Remark 10.3.5. The main result of Theorem 10.3.2 can be viewed as a property of *global exponential stability* of the positive stationary solution $\varphi(x)$ in the metric space $Z \setminus \{0\} = \{u \in C_0(\Omega);\ u \geq 0,\ u \neq 0\}$. Here, three remarks are in order.

1. The constant $C(u_0)$ in (10.35) does not remain bounded with $\|u_0\|_{L^\infty}$. In fact, let $A > 0$ arbitrary and select $t = T$ such that $\exp(\gamma T)\|\varphi\|_{L^\infty} > A$. By letting $u_0 \to 0$ in $C_0(\Omega)$, we deduce from (10.5) with $t = T$ the estimate

$$\liminf\{C(u_0),\ u_0 \to 0 \text{ in } C_0(\Omega)\} > A.$$

Since A is arbitrary, we conclude: $\lim\{C(u_0),\ u_0 \to 0 \text{ in } C_0(\Omega)\} = \infty$.

2. Assuming that $f : \mathbb{R} \to \mathbb{R}$ is *odd*, locally Lipschitz continuous with $f(s) \geq 0$ for $s \to \infty$, and satisfies (10.19), it also follows from the proof of Theorem 10.3.2 that the positive stationary solution $\varphi(x)$ is *exponentially stable* in the larger space $C_0(\Omega)$. Indeed, the linearized equation around $u = \varphi(x)$ is

$$\begin{cases} z_t - \Delta z + f'(\varphi)z = 0 & \text{in } \mathbb{R}^+ \times \Omega, \\ z = 0 & \text{on } \mathbb{R}^+ \times \partial\Omega \end{cases} \quad (10.36)$$

which, using the convexity of f on $\mathbb{R}^+$, turns out to be exponentially damped in $C_0(\Omega)$ by the method of Lemma 10.3.4.

3. Theorem 10.3.2 and the two remarks above are applicable, as a typical case, to the non-linearity

$$f(u) = c|u|^\alpha u - \lambda u \quad (10.37)$$

for some positive constants c, α, and λ.

(c) Similarly, we can consider the semilinear wave equation

$$u_{tt} - \Delta u + f(u) + \lambda u_t = 0 \quad \text{in } \mathbb{R}^+ \times \Omega;\quad u = 0 \quad \text{on } \mathbb{R}^+ \times \partial\Omega \quad (10.38)$$

where Ω is a bounded domain in $\mathbb{R}^N$, f is a function of class $C^1 : \mathbb{R} \to \mathbb{R}$ such that

$$f(0) = 0 \quad \text{and} \quad f'(0) > -\lambda_1, \quad (10.39)$$

in which f is a locally Lipschitz continuous function: $\mathbb{R} \to \mathbb{R}$ with $f(0) = 0$ satisfying the growth condition

$$|f(u)| \leq C(1 + |u|^r) \quad \text{a.e. on } \mathbb{R} \quad (10.40)$$

with $r \geq 0$ arbitrary if $N = 1$ or 2 and $0 \leq r \leq N/(N-2)$ if $N \geq 3$.

We obtain the following result.

Proposition 10.3.6. Under the above hypotheses, the stationary solution $(u, v) \equiv (0, 0)$ of (10.38) is exponentially stable in $X = H_0^1(\Omega) \times L^2(\Omega)$ in the following sense: there exist $\delta > 0$ and $R = R(\delta)$ such that, for all $x \in X$ with $\|x\| \leq R$, the solution u of (10.38) such that $u(0) = x$ is global and satisfies

$$\forall t \geq 0, \quad \|u(t)\| \leq M(\delta)\|x\|\exp(-\delta t). \tag{10.41}$$

Proof. It is well known, and this can be deduced from the proof of Theorem 8.4.5 and Lemma 8.4.6 with $H = 0$, that the contraction semigroup $T_0(t)$ generated in $X = H_0^1(\Omega) \times L^2(\Omega)$ by the equation

$$\begin{cases} u_{tt} - \Delta u + \lambda u_t = 0 & \text{in } \mathbb{R}^+ \times \Omega, \\ u = 0 & \text{on } \mathbb{R}^+ \times \partial \Omega, \end{cases}$$

satisfies (10.10). The method of proof clearly applies to the slightly more general equation

$$\begin{cases} u_{tt} - \Delta u + f'(0)u + \lambda u_t = 0 & \text{in } \mathbb{R}^+ \times \Omega, \\ u = 0 & \text{on } \mathbb{R}^+ \times \partial \Omega. \end{cases} \tag{10.42}$$

In order to apply Theorem 10.2.2 with $T(t)$ the semigroup generated by (10.42), all we need to check is that the function $F(u, v) = -(0, f(u) - f'(0)u)$ satisfies (10.11) with $a = 0$ and ν arbitrarily small. But this is immediate, since the function $\varphi(s) = f(s) - f'(0)s$ is $o(|s|)$ near the origin and, by (10.40), we have $|\varphi(s)| \leq C(|s|^r)$ for s large. Therefore, for each δ arbitrarily small, we have $|\varphi(s)| \leq \delta|s| + C(\delta)|s|^r$, globally on $\mathbb{R}$. The result then follows immediately from the Sobolev embedding theorems. □

10.4. Stability and positivity

We consider the semilinear parabolic equation (10.16), where f is a locally Lipschitz continuous function on the reals. If u is solution of (10.16) uniformly bounded on $\mathbb{R}^+ \times \Omega$, the solution $u(t,.)$ asymptotes towards the set of stationary solutions as $t \to \infty$. In particular, the existence of a bounded trajectory implies the existence of at least one solution to the elliptic problem (10.20). Roughly speaking, the stability of a solution φ of (10.20) rests on the sign of the first eigenvalue $\eta = \lambda_1(-\Delta + f'(\varphi)I)$ in the sense of $H_0^1(\Omega)$; more precisely, if $\eta > 0$, φ is Liapunov-stable in the uniform norm as well as in any reasonable usual norm, while if $\eta < 0$, φ is not Liapunov-stable. In the more specific case in which f is *strictly convex* on $[0, \infty)$ with $f(0) = 0$ and $f'_d(0) < -\lambda_1(\Omega)$, where $\lambda_1(\Omega)$ is the smallest eigenvalue of $(-\Delta)$ in $H_0^1(\Omega)$, we have proved in §10.3 that the unique *positive solution* φ is in fact exponentially stable in the space $C_0(\Omega)$. This is because the strict convexity of f implies that $\eta = \lambda_1(-\Delta + f'(\varphi)I) > 0$.

A typical example where the exponential stability of the positive solution occurs is when
$$f(u) = c|u|^\alpha u - \lambda u$$
for some constants c, $\alpha > 0$ and $\lambda > \lambda_1(\Omega)$. When $N = 1$, $\Omega = (0, L)$, and for $\lambda > (\pi/2)^2 = \lambda_1(\Omega)$, non-trivial stationary solutions appear as pairs of opposite functions that have a finite number $(n + 2)$ of zeroes equally spaced on $[0, L]$ with $n \leq (L/\pi)\lambda^{1/2} - 1$, built from positive solutions of the same problem on $(0, L/(n+1))$. A new pair of solutions appears when the increasing positive parameter λ crosses an eigenvalue $(k\pi/L)^2 = \lambda_k(0, L)$ of the operator $(-u_{xx})$ in $H_0^1(0, L)$. It has been known for some time (cf., e.g., Chafee and Infante [1]) that the only stationary solutions that are stable in the sense of Liapunov are the positive and the negative solution. In higher dimensions the situation seems much more intricate, but it is still of interest to investigate the relationship between stability and the absence of zeroes. For instance, in the case of Neumann boundary conditions, a result of Casten and Holland [1] asserts that if Ω is star-shaped, any non-constant solution is unstable. Counter-examples of stable solutions changing sign in Ω are known for both Dirichlet and Neumann boundary conditions in non-convex domains. On the other hand, even for Ω convex, there is no general instability result for solutions changing sign in the case of Dirichlet boundary conditions.

10.4.1. The one-dimensional case

Consider, as a motivation, the one-dimensional semilinear heat equation

$$\begin{cases} u_t - u_{xx} + f(u) = 0 & \text{in } \mathbb{R}^+ \times (0, L), \\ u(t, 0) = u(t, L) = 0 & \text{on } \mathbb{R}^+, \end{cases} \quad (10.43)$$

where f is a C^1 function: $\mathbb{R} \to \mathbb{R}$. Any solution u of this problem which is global and uniformly bounded on $\mathbb{R}^+ \times (0, L)$ converges as $t \to \infty$ to a solution φ of the elliptic problem

$$\varphi \in H_0^1(0, L), \quad -\varphi_{xx} + f(\varphi) = 0. \quad (10.44)$$

Proposition 10.4.1. *If φ is a stable solution of (10.44), then φ has a constant sign on $(0, L)$.*

Proof. Indeed, if φ is not identically 0 and vanishes somewhere in $(0, L)$, the function $w := \varphi'$ has two zeroes in $(0, L)$ and satisfies

$$w \in C^2([0, L]) \cap H_0^1(0, L), \quad -w_{xx} + f'(\varphi)w = 0 \quad \text{in } (0, L).$$

In particular, if $0 < \alpha < \beta < L$ are such that $w(\alpha) = w(\beta) = 0$, $w \neq 0$ on (α, β) and if we set $\omega = (\alpha, \beta)$, we clearly have $\lambda_1(\omega; -\Delta + f'(\varphi)I) = 0$, where $\lambda_1(\omega; -\Delta + f'(\varphi)I)$ denotes the first eigenvalue of $-\Delta + f'(\varphi)I$ in the sense of $H_0^1(\omega)$. We introduce the quadratic form J defined by

$$\forall z \in H_0^1(0, L) \quad J(z) := \int_\Omega \{z_x^2 + f'(\varphi)z^2\} dx.$$

Let

$$\eta := \inf\{J(z), \ z \in H_0^1(0, L), \ \int_\Omega z^2 \, dx = 1\}.$$

Let us also denote, by ζ, the extension of w by 0 outside ω. Because $J(\zeta) = \int_\Omega \{\zeta_x^2 + f'(\varphi)\zeta^2\} dx = \int_\omega \{\zeta_x^2 + f'(\varphi)\zeta^2\} dx = \int_\omega \{w_x^2 + f'(\varphi)w^2\} dx = 0$, we clearly have

$$\eta = \lambda_1(\Omega; -\Delta + f'(\varphi)I) \leq 0.$$

Assuming that $\eta = 0$ means that a multiple $\lambda\zeta = \psi$ of ζ realizes the minimum of J and therefore is a solution of

$$\psi \in C^2([0, L]) \cap H_0^1(0, L), \quad -\psi_{xx} + f'(\varphi)\psi = 0.$$

This is impossible since ψ is not identically 0 and yet vanishes on $(0, \alpha)$. Therefore $\eta < 0$. The deduction of instability will now follow in a few lines: assume that φ is stable in the sense of Liapunov in $C_0(\Omega) \times H_0^1(\Omega)$ and let u_ε be the solution of (10.43) with $u_\varepsilon(0) = \varphi + \varepsilon\psi$, where ψ is a positive solution of

$$\psi \in C^2 \cap H_0^1([0, L]), \quad -\psi_{xx} + f'(\varphi)\psi = \eta\psi \quad \text{in } (0, L).$$

Since $\psi > 0$, the order preserving property implies that $u_\varepsilon \geq \varphi$. Now let $w_\varepsilon = u_\varepsilon - \varphi \geq 0$.

Because $w_\varepsilon \to 0$ uniformly as $\varepsilon \to 0$, we have

$$f(u_\varepsilon) = f(\varphi) + f'(\varphi)w_\varepsilon + \delta(\varepsilon)|w_\varepsilon|,$$

with $\delta(\varepsilon) \to 0$ as $\varepsilon \to 0$. On the other hand, we have

$$\frac{d}{dt}\int_\Omega w_\varepsilon \psi(x) \, dx = \int_\Omega \psi(u_{\varepsilon xx} - f(u_\varepsilon)) dx$$

$$= \int_\Omega \psi\varphi_{xx} + \int_\Omega \psi_{xx} w_\varepsilon \, dx - \int_\Omega f(\varphi)\psi \, dx$$

$$- \int_\Omega f'(\varphi)w_\varepsilon \psi \, dx - \int_\Omega \delta(\varepsilon)|w_\varepsilon|\psi \, dx$$

$$= -\eta \int_\Omega w_\varepsilon \psi(x) \, dx - \int_\Omega \delta(\varepsilon)|w_\varepsilon|\psi \, dx$$

$$\geq (|\eta|/2) \int_\Omega w_\varepsilon \psi(x) \, dx,$$

for all $t \geq 0$ and $\varepsilon > 0$ small enough. This inequality, combined with positivity of ψ and w_ε together with boundedness of w_ε, implies that

$$0 = \int_\Omega \psi(x) w_\varepsilon(0,x) \mathrm{d}x = \varepsilon \int_\Omega \psi^2(x) \mathrm{d}x.$$

Thus Proposition 10.4.1 is proven by contradiction. □

10.4.2. The multidimensional case

It is interesting to remark that the above instability result does not require hypotheses on the shape of f. Even the differentiability condition can in fact be relaxed, since the important point is just boundedness of the potential $f'(\varphi(x))$. In higher dimensions, at present we have no such general results; however, the previous technique can be extended to some particular cases of special interest. First, we establish a basic lemma.

Lemma 10.4.2. *Let Ω be any bounded open domain of $\mathbb{R}^N$, and let ω be any open sub-domain which is not dense in Ω. Then, for any potential $p \in L^\infty(\Omega)$, we have*

$$\lambda_1(\Omega;\; -\Delta + p(x)I) < \lambda_1(\omega;\; -\Delta + p(x)I).$$

Proof. We introduce the quadratic form J, defined by

$$\forall z \in H_0^1(\Omega), \quad J(z) := \int_\Omega \{|\nabla z|^2 + p(x)z^2\} \mathrm{d}x.$$

By definition,

$$\eta := \inf\{J(z),\; z \in H_0^1(0,L),\; \int_\Omega z^2\, \mathrm{d}x = 1\} = \lambda_1(\Omega;\; -\Delta + p(x)I).$$

Let $w \in H_0^1(\omega)$ denote a normalized eigenvector of $-\Delta + p(x)I$ in $H_0^1(\omega)$ associated with $\lambda_1(\omega;\; -\Delta + p(x)I)$ and let us consider the extension $\zeta \in H_0^1(\Omega)$ of w by 0 outside ω. Since

$$J(\zeta) = \int_\Omega \{|\nabla \zeta|^2 + p(x)\zeta^2\} \mathrm{d}x = \int_\omega \{|\nabla \zeta|^2 + p(x)\zeta^2\} \mathrm{d}x$$
$$= \int_\omega \{|\nabla w|^2 + p(x)w^2\} \mathrm{d}x = \lambda_1(\omega;\; -\Delta + p(x)I),$$

we clearly have $\eta \leq \lambda_1(\omega;\; -\Delta + p(x)I)$.

Assuming that $\eta = \lambda_1(\omega;\; -\Delta + p(x)I)$ means that a real multiple of ζ realizes the minimum of J and therefore $-\Delta \zeta + p(x)\zeta = 0$. This is impossible, since ζ is not identically 0 and yet vanishes on a non-empty open subset of Ω. □

An especially interesting special case is when Ω is a rectangle in $\mathbb{R}^2$, and if f is a C^1 odd function on the reals. For instance, if $f(u) = cu^3 - \lambda u$ with λ large, in addition to the unique positive solution, (10.20) will have more complicated solutions arising by odd extensions from the positive solutions in cellular subdomains which divide the domain into $m \times p$ equal rectangular regions. Such solutions, which are the analogue of odd-periodic extensions in the one-dimensional case, are also Liapunov-unstable. More precisely, we have the following.

Proposition 10.4.3. *Let φ be a solution of (10.20) in a rectangle $\Omega = (0, a) \times (0, b)$ and assume that there is a sub-rectangle $R = (0, a/p) \times (0, b/q)$ with p, q integers ≥ 1 and $\max(p, q) > 1$ such that φ has a constant sign in R and $\varphi = 0$ on ∂R. Then φ is unstable.*

Proof. Assume, for instance, that $p > 1$. It is clear that the trace of φ on $R' = (a/p, 2a/p) \times (0, b/q)$ coincides with the odd reflection of $\varphi_{|R}$ with respect to the line $x = a/p$, and in addition $\varphi_{|R}$ is even with respect to the line $x = a/2p$. Therefore, $w := \partial \varphi / \partial x$ vanishes on the boundary of the sub-rectangle $\omega = (a/2p, 3a/2p) \times (0, b/q)$ and satisfies

$$w \in C^2(\omega) \cap H_0^1(\omega), \quad -\Delta w + f'(\varphi)w = 0 \quad \text{in } \omega.$$

In particular, since w does not vanish inside ω, we have $\lambda_1(\omega; -\Delta + f'(\varphi)I) = 0$. Then Lemma 10.4.2 gives

$$\lambda_1(\Omega; -\Delta + f'(\varphi)I) < \lambda_1(\omega; -\Delta + f'(\varphi)I) = 0.$$

Therefore φ is unstable. $\square$

There is another quite interesting result for the case in which Ω is a sphere in $\mathbb{R}^N$: in Comte, Haraux, and Mironescu [1] the following result is obtained.

Proposition 10.4.4. *If Ω is a ball in $\mathbb{R}^N$, $N \geq 2$, and φ is a Liapunov-stable solution of (10.20), we have either $\varphi = 0$, or φ is spherically symmetric with constant sign in Ω.*

Notes. Theorem 10.2.2 is also valid when $(\mathcal{T}(t))_{t \geq 0}$ is a general C_0-semigroup; see Haraux [5]. The exponential stability property (10.21) can also be established in a non-autonomous framework; see Haraux [6].

Further remarks. We have not addressed the Navier–Stokes equation. The reader may consult Constantin and Foias [1], Foias and Temam [1], Fujita and Kato [1], Giga [1, 2], Heywood [1], Ito [1], Kato [1], Ladyzhenskaya [1], Leray [1–3], Temam [1], and Lions [3, 4]. For the Korteweg–De Vries equation, see, for example, Gardner, Greene, Kruskal, and Miura [1], and Kato [2].

Bibliography

ADAMS R. A.
[1] *Sobolev spaces*, Academic Press, New York, 1975.

AGUIRRE J. and ESCOBEDO M.
[1] A Cauchy problem for $u_t - \Delta u = u^p$ with $0 < p < 1$. Asymptotic behavior of solutions, Ann Fac. Sci. Toulouse **8** (1986–87), 175–203.

BABIN A. V. and VISHIK M. I.
[1] Regular attractors of semi-groups and evolution equations, J. Math. Pures Appl. **62** (1983), 441–491.
[2] Attractors of some evolution equations and estimates of their dimensions, Uspekhi Math. Nauk. **38** (1983), 133–187 (in Russian).
[3] Attracteurs maximaux dans les équations aux dérivées partielles, in *Nonlinear partial differential equations and their applications, College de France Seminar, vol. 7*, H. Brezis and J.L. Lions (eds), Research Notes in Math. #**122**, Pitman, 1984, 1–34.

BAILLON J.-B. and CHADAM J. M.
[1] The Cauchy problem for the coupled Schrödinger–Klein–Gordon equations, in *Contemporary developments in continuum mechanics and partial differential equations*, North Holland Math. Studies, North Holland, 1978.

BAILLON J.-B., CAZENAVE T., and FIGUEIRA M.
[1] Equation de Schrödinger non linéaire, C. R. Acad. Sci. Paris **284** (1977), 869–872.
[2] Equation de Schrödinger avec non-linéarité intégrale, C. R. Acad. Sci. Paris **284** (1977), 939–942.

BALABANE M.
[1] Non existence de solutions globales pour des équations des ondes non linéaires à données de Cauchy petites, C. R. Acad. Sci. Paris **301** (1985), 569–572.
[2] Ondes progressives et résultats d'explosion pour les systèmes du premier ordre, C. R. Acad. Sci. Paris **302** (1986), 211–214.

BALL J. M.
[1] On the asymptotic behavior of generalized processes, with applications to nonlinear evolution equations, J. Diff. Eq. **27** (1978), 224–265.

[2] Finite-time blow-up in nonlinear problems, in *Nonlinear evolution equations*, M. G. Crandall (ed.), Academic Press, New York, 1978, 189–205.
[3] Remarks on blow-up and nonexistence theorems for nonlinear evolution equations, Q. J. Math. Oxford **28** (1977), 473–486.

BARAS P.
[1] Non unicité des solutions d'une équation d'évolution non linéaire, Ann. Fac. Sci. Toulouse **5** (1983), 287–302.

BARAS P. and COHEN L.
[1] Complete blow-up after T_{max} for the solution of a semilinear heat equation, J. Funct. Anal. **71** (1987), 142–174.

BARAS P. and GOLDSTEIN J.
[1] The heat equation with a singular potential, Trans. Am. Math. Soc. **284** (1984), 121–139.

BERESTYCKI H. and CAZENAVE T.
[1] Instabilité des états stationnaires dans les équations de Schrödinger et de Klein–Gordon non linéaires, C. R. Acad. Sci. Paris **293** (1981), 489–492.

BERESTYCKI H. and LIONS P.-L.
[1] Nonlinear scalar field equations, Arch. Rat. Mech. Anal. **82** (1983), 313–375.

BERESTYCKI H., GALLOUET T., and KAVIAN O.
[1] Équations de champs scalaires Euclidiens non-linéaires dans le plan, C. R. Acad. Sci. Paris **297** (1983), 307–310.

BERESTYCKI H., LIONS P.-L., and PELETIER L. A.
[1] An O.D.E. approach to the existence of positive solutions for semilinear problems in $\mathbb{R}^n$, Indiana Univ. Math. J. **30** (1981), 141–157.

BERGH J. and LÖFSTRÖM J.
[1] *Interpolation spaces*, Springer, New York, 1976.

BLANCHARD P., STUBBE J. and VÁZQUEZ L.
[1] On the stability of solitary waves for classical scalar fields, Ann. Inst. Henri Poincaré, Physique Théorique **47** (1987), 309–336.

BRENNER P.
[1] On $L^p - L^{p'}$ estimates for the wave equation, Math. Z. **145** (1975), 251–254.
[2] On space–time means and everywhere defined scattering operators for nonlinear Klein–Gordon equations, Math. Z. **186** (1984), 383–391.

BREZIS H.
[1] *Opérateurs maximaux monotones et semi-groupes de contractions dans les espaces de Hilbert*, Lecture Notes #5, North–Holland, Amsterdam, London, 1972.
[2] *Analyse fonctionnelle, théorie et applications*, Masson, Paris, 1983.

BREZIS H. and FRIEDMAN A.
[1] Nonlinear parabolic equations involving measures as initial conditions, J. Math. Pures Appl. **62** (1983), 73–97.

BREZIS H. and GALLOUET T.
[1] Nonlinear Schrödinger evolution equations, Nonlinear Anal. T.M.A. **4** (1980), 677–681.

BREZIS H., CORON J.-M., and NIRENBERG L.
[1] Free vibrations for a nonlinear wave equation and a theorem of P. Rabinowitz, Communs. Pure Appl. Math. **33** (1980), 667–689.

BREZIS H., PELETIER L. A., and TERMAN D.
[1] A very singular solution of the heat equation with absorption, Arch. Rat. Mech. Anal. **95** (1986), 185–209.

BROWDER F.
[1] On nonlinear wave equations, Math. Z. **80** (1962), 249–264.

CABANNES H. and HARAUX A.
[1] Mouvements presque-périodiques d'une corde vibrante en présence d'un obstacle fixe, rectiligne ou ponctuel, Int. J. Nonlinear Mech. **16** (1981), 449–458.
[2] Almost periodic motion of a string vibrating against a straight, fixed obstacle, Nonlinear Anal. T.M.A. **7** (1983), 129–141.

CASTEN R. J. and HOLLAND C. J.
[1] Instability results for reaction diffusion equations with Neumann boundary conditions, J. Diff. Eq. **27** (1978), 266–273.

CAZENAVE T.
[1] Uniform estimates for solutions of nonlinear Klein–Gordon equations, J. Funct. Anal. **60** (1985), 36–55.
[2] Equations de Schrödinger non linéaires en dimension deux, Proc. R. Soc. Edin. **84** (1979), 327–346.
[3] Stable solutions of the logarithmic Schrödinger equation, Nonlinear Anal. T. M. A. **7** (1983), 1127–1140.
[4] *An introduction to nonlinear Schrödinger equations*, Textos de Métodos Matemáticos #**22**, I.M.U.F.R.J., Rio de Janeiro, 1989.

CAZENAVE T. and HARAUX A.
[1] Équation d'évolution avec non-linéarité logarithmique, Ann. Fac. Sci. Toulouse **2** (1980), 21–51.
[2] Oscillatory phenomena associated to semilinear wave equations in one spatial dimension, Trans. Am. Math. Soc. **300** (1987), 207–233.
[3] Some oscillation properties of the wave equation in several space dimensions, J. Funct. Anal. **76** (1988), 87–109.

[4] *Introduction aux problèmes d'évolution semi-linéaires*, Mathématiques et Applications #1, Ellipses, Paris, 1990.

CAZENAVE T. and LIONS P.-L.
[1] Orbital stability of standing waves for some nonlinear Schrödinger equations, Communs. Math. Phys. **85** (1982), 549–561.
[2] Solutions globales d'équations de la chaleur semi lineaires, Communs. Part. Diff. Eq. **9** (1984), 955–978.

CAZENAVE T. and WEISSLER F. B.
[1] Some remarks on the nonlinear Schrödinger equation in the subcritical case, in *New methods and results in nonlinear field equations*, Ph. Blanchard, J.-P. Dias, and J. Stubbe (eds), Lect. Notes in Phys. #**347**, Springer, 1989, 59–69.
[2] Some remarks on the nonlinear Schrödinger equation in the critical case, in *Nonlinear semigroups, partial differential equations, and attractors*, T. L. Gill and W. W. Zachary (eds), Lect. Notes in Math. # **1394**, Springer, 1989, 18–29.
[3] The Cauchy problem for the nonlinear Schrödinger equation in H^1, Manuscripta Math. **61** (1988), 477–494.
[4] The Cauchy problem for the critical nonlinear Schrödinger equation in H^s, Nonlinear Anal, T.M.A. **14** (1990), 807–836.
[5] The structure of solutions to the pseudo-conformally invariant nonlinear Schrödinger equation, Proc. R. Soc. Edin. **117A** (1991), 251–273.
[6] Rapidly decaying solutions of the nonlinear Schrödinger equation, Communs. Math. Phys. **147** (1992), 75–100.

CAZENAVE T., HARAUX A., VÁZQUEZ L., and WEISSLER F. B.
[1] Nonlinear effects in the wave equation with a cubic restoring force, Comput. Mech. **5** (1989), 49–72.

CHADAM J. M. and GLASSEY R. T.
[1] Global existence of solutions to the Cauchy problem for time-dependent Hartree equations, J. Math. Phys. **16** (1975), 1122–1130.

CHAFEE N. and INFANTE L. I.
[1] A bifurcation problem for a nonlinear partial differential equation of parabolic type, Applic. Anal. **4** (1974), 17–37.

COMTE M., HARAUX A., and MIRONESCU P.
[1] Multiplicity and stability topics in semilinear parabolic problems, Publ. Lab. Anal. Numérique R96038 (1991).

CONSTANTIN P. and FOIAS C.
[1] *Navier–Stokes equations*, Chicago University Press, 1989.

CONSTANTIN P. and SAUT J.-C.
[1] Effets régularisant locaux pour des équations dispersives générales, C. R. Acad. Sci. Paris **304** (1987), 407–410.

[2] Local smoothing properties of dispersive equations, J. Am. Math. Soc. **1** (1988), 413–439.
[3] Local smoothing properties of Schrödinger equations, Indiana Univ. Math. J. **38** (1989), 791–810.

COURANT R. and HILBERT D.
[1] *Methods of mathematical physics*, vol. *2*, Interscience, John Wiley, 1962.

CRANDALL M. G. and LIGGETT T.
[1] Generation of semigroups of nonlinear transformations on general Banach spaces, Am. J. Math. **93** (1971), 265–298.

CRANDALL M. G. and PAZY A.
[1] Semigroups of nonlinear contractions and dissipative sets, J. Funct. Anal. **3** (1969), 376–418.

DAFERMOS C. M.
[1] *Contraction semigroups and trend to equilibrium in continuum mechanics*, Lect. Notes in Math. #**503**, Springer, 1976, 295–306.
[2] Almost periodic processes and almost periodic solutions of evolution equations, Proceedings of a University of Florida International Symposium, Academic Press, New York, 1977, 43–57.
[3] Asymptotic behavior of solutions of evolution equations, in *Nonlinear evolution equations*, M. G. Crandall (ed.), Academic Press, New York, 1978, 103–123.

DAFERMOS C. M. and SLEMROD M.
[1] Asymptotic behavior of nonlinear contraction semi-groups, J. Funct. Anal. **12** (1973), 97–103.

DIAS J.-P. and FIGUEIRA M.
[1] Conservation laws and time decay for the solutions to some nonlinear Schrödinger–Hartree equations, J. Math. Anal. Appl. **84** (1981), 486–508.

DIAS J.-P. and HARAUX A.
[1] Smoothing effect and asymptotic behavior for the solutions of a nonlinear time dependent system, Proc. R. Soc. Edin. **87** (1981), 289–303.

DINCULEANU N.
[1] *Vector measures*, Pergamon Press, New York, 1967.

DUNFORD C. and SCHWARTZ J. T.
[1] *Linear operators*, Interscience, New York, 1958.

ESCOBEDO M. and KAVIAN O.
[1] Variational problems related to self-similar solutions of the heat equation, Nonlinear Anal. T.M.A. **11** (1987), 1103–1133.

[2] Asymptotic behavior of positive solutions of a nonlinear heat equation, Houston J. Math. **13** (1987), 39–50.

ESTEBAN M. J.
[1] On periodic solutions of superlinear parabolic problems, Trans. Am. Math. Soc **293** (1986), 171–189.
[1] A remark on the existence of positive periodic solutions of superlinear parabolic problems, Proc. Am. Math. Soc. **102** (1987), 131–136.

FIFE P. C.
[1] *Mathematical aspects of reacting and diffusing systems*, Lecture Notes in Biomathematics #**28**, Springer, New York, 1979.

FOIAS C. and TEMAM R.
[1] Some analytic and geometric properties of the solutions of the Navier–Stokes equations, J. Math. Pure Appl. **58** (1979), 339–368.

FRIEDLANDER L.
[1] An invariant measure for the equation $u_{tt} - u_{xx} + u^3 = 0$, Communs. Math. Phys. **98** (1985), 1–16.

FRIEDMAN A.
[1] *Partial differential equations of parabolic type*, Prentice-Hall, 1964.

FRIEDMAN A. and GIGA Y.
[1] A single point blow-up for solutions of semilinear parabolic equations, J. Fac. Sci. Univ. Tokyo, Sect. IA, **34** (1987), 65–79.

FRIEDMAN A. and McLEOD B.
[1] Blow-up of positive solutions of semilinear heat equations, Indiana Univ. Math. J. **34** (1985), 425–447.

FUJITA H.
[1] On the blowing-up of solutions of the Cauchy problem for $u_t = \Delta u + u^{\alpha+1}$, J. Fac. Sci. Univ. Tokyo **13** (1966), 109–124.

FUJITA H. and KATO T.
[1] On the Navier–Stokes initial value problem, Arch. Rat. Mech. Anal. **16** (1964), 269–315.

GARDNER C. S., GREENE J., KRUSKAL M. D., and MIURA R. M.
[1] Korteweg–De Vries equation and generalizations, methods for exact solutions, Communs. Pure Appl. Math. **28** (1974), 97–133.

GHIDAGLIA J.-M. and TEMAM R.
[1] Propriétés des attracteurs associés à des équations des ondes amorties, C. R. Acad. Sci. Paris **300** (1985), 185–188.

[2] Attractors for damped nonlinear hyperbolic equations, J. Math. Pures Appl. **66** (1987), 233–319.

GIGA Y.
[1] The Navier–Stokes initial value problem in L^p and related problems, in Lecture Notes in Num. Appl. Anal., 5 (1982), 37–54.
[2] Weak and strong solutions of the Navier–Stokes initial value problem, Publ. Res. Inst. Math. Sci. (Kyoto Univ.) **19** (1983), 887–910.
[3] A bound for global solutions of semilinear heat equations, Communs. Math. Phys. **103** (1986), 415–421.

GIGA Y. and KOHN R. V.
[1] Asymptotically self-similar blow-up of semilinear heat equations, Communs. Pure Appl. Math. **38** (1985), 297–319.
[2] Characterizing blow up using similarity variables, Indiana Univ. Math. J. **36** (1987), 1–40.
[3] Nondegeneracy of blowup for semilinear heat equations, Communs. Pure Appl. Math. **62** (1989), 845–885.
[4] Removability of blowup points for semilinear heat equations, in *Differential equations, Xanthi, 1987*, G. Papanicolaou (ed.), Lecture Notes in Pure and Appl. Math. **118**, Marcel Dekker, New York, 1989, 257–264.

GILBARG D. and TRUDINGER N. S.
[1] *Elliptic partial differential equations of the second order*, Springer-Verlag, Berlin, 1983.

GINIBRE J. and VELO G.
[1] On a class of nonlinear Schrödinger equations, J. Funct. Anal. **32** (1979), 1–71.
[2] On a class of nonlinear Schrödinger equations, special theories in dimensions 1, 2 and 3, Ann. Inst. Henri Poincaré **28** (1978), 287–316.
[3] On a class of nonlinear Schrödinger equations with non local interaction, Math. Z. **170** (1980), 109–136.
[4] On the global Cauchy problem for some nonlinear Schrödinger equations, Ann. Inst. Henri Poincaré, Analyse Non Linéaire **1** (1984), 309–323.
[5] The global Cauchy problem for the nonlinear Schrödinger equation revisited, Ann. Inst. Henri Poincaré, Analyse Non Linéaire **2** (1985), 309–327.
[6] The global Cauchy problem for the nonlinear Klein–Gordon equation, Math. Z. **189** (1985), 487–505.
[7] Scattering theory in the energy space for a class of nonlinear Schrödinger equations, J. Math. Pure Appl. **64** (1985), 363–401.
[8] Time decay of finite energy solutions of the nonlinear Klein–Gordon and Schrödinger equations, Ann. Inst. Henri Poincaré, Physique Théorique **43** (1985), 399–442.

[9] The global Cauchy problem for the nonlinear Klein–Gordon equation II, Ann. Inst. Henri Poincaré, Analyse Non Linéaire **6** (1989), 15–35.
[10] Conformal invariance and time decay for nonlinear wave equations, Ann. Inst. Henri Poincaré, Physique Théorique **47** (1987), 221–276.
[11] Sur une équation de Schrödinger non-linéaire avec interaction non-locale, in *Nonlinear partial differential equations and their applications, College de France Seminar, vol. 2*. H. Brezis and J.L. Lions (eds), Research Notes in Math. #60, Pitman, 1982, 155–199.
[12] Scattering theory in the energy space for a class of nonlinear wave equations, Communs. Math. Phys. **123** (1989), 535–573.

GLASSEY R. T.
[1] Blow up theorems for nonlinear wave equations, Math. Z. **132** (1973), 183–203.
[2] On the blowing up of solutions to the Cauchy problem for nonlinear Schrödinger equations, J. Math. Phys. **18** (1977), 1794–1797.
[3] Finite-time blow up for solutions of nonlinear wave equations, Math. Z. **177** (1981), 323–340.

GOLDSTEIN J.
[1] *Semigroups of linear operators and applications to partial differential equations*, Oxford University Press, New York, 1986.

GRILLAKIS M., SHATAH J. and STRAUSS W. A.
[1] Stability theory of solitary waves in the presence of symmetry I, J. Funct. Anal. **74** (1987), 160–197.

HALE J. K.
[1] *Ordinary differential equations*, Wiley-Interscience, New York, 1969.
[2] Asymptotic behavior and dynamics in infinite dimensions, in *Nonlinear differential equations*, Hale J. K. and Martines-Amores P. (eds), Research Notes in Math. #132, Pitman, 1985, 1–42.

HALE J. K. and RAUGEL G.
Convergence in gradient-like systems with applications to PDE, Z. Math. Phys. **43** (1992), 63–124.

HANOUZET B. and JOLY J.-L.
[1] Explosion pour des problèmes hyperboliques semi linéaires avec second membre compatible, C. R. Acad. Sci. Paris **301** (1985), 581–584.

HARAUX A.
[1] *Nonlinear evolution equations: global behavior of solutions*, Lecture Notes in Math. #841, Springer, 1981.
[2] *Semi-linear wave equations in bounded domains*, Mathematical reports, J. Dieudonné (ed.), Vol. 3, Part 1, Harwood, London, 1987.

[3] Linear semi-groups in Banach spaces, in *Semigroups, theory and applications*, H. Brezis, M. G. Crandall, and F. Kappel (eds), Pitman Research Notes in Mathematics series #**152**, Longman, 1986, 93–135.
[4] Nonlinear vibrations and the wave equation, Textos de Metodos Matematicos #**20**, I.M.U.F.R.J., Rio de Janeiro, 1987.
[5] *Systèmes dynamiques dissipatifs et applications*, R.M.A.#**17**, P. G. Ciarlet and J.-L. Lions (eds), Masson, Paris, 1991.
[6] Exponentially stable positive solutions to a forced semilinear parabolic equation, Asymptotic Anal. **7** (1993), 3–13.

HARAUX A. and KIRANE M.
[1] Estimations C^1 pour des problèmes paraboliques semi-linéaires, Ann. Fac. Sci. Toulouse **5** (1983), 265–280.

HARAUX A. and POLÁČIK P.
[1] Convergence to a positive equilibrium for some semilinear evolution equations in a ball, Acta. Math. Univ Comenianae LXI **2** (1993), 129–141.

HARAUX A. and WEISSLER F. B.
[1] Non uniqueness for a semilinear initial value problem, Indiana Univ. Math. J. **31** (1982), 167–189.

HARAUX A. and YOUKANA A.
[1] On a result of K. Masuda concerning reaction–diffusion equations. Tôhoku Math. J. **40** (1988), 159–163.

HAYASHI N.
[1] Classical solutions of nonlinear Schrödinger equations, Manuscripta Math. **55** (1986), 171–190.
[2] Asymptotic behavior of solutions to time-dependent Hartree equations, Arch. Rat. Mech. Anal. **100** (1988), 191–206.

HAYASHI N. and OZAWA T.
[1] Time decay of solutions to the Cauchy problem for time-dependent Schrödinger–Hartree equations, Communs. Math. Phys **110** (1987), 467–478.
[2] Smoothing effect for some Schrödinger equations, J. Funct. Anal. **85** (1989), 307–348.

HAYASHI N. and TSUTSUMI M.
[1] Classical solutions of nonlinear Schrödinger equations in higher dimensions, Math. Z. **177** (1981), 217–234.

HAYASHI N. and TSUTSUMI Y.
[1] Scattering theory for Hartree type equations, Ann. Inst. Henri Poincaré, Physique Théorique **46** (1987), 187–213.

HAYASHI N., NAKAMITSU K., and TSUTSUMI M.
[1] On solutions of the initial value problem for the nonlinear Schrödinger equations in one space dimension, Math. Z. **192** (1986), 637–650.
[2] On solutions of the initial value problem for the nonlinear Schrödinger equations, J. Funct. Anal. **71** (1987), 218–245.
[3] Nonlinear Schrödinger equations in weighted Sobolev spaces, Funk. Ekva **31** (1988), 363–381.

HEINZ E. and VON WAHL W.
[1] Zu einem Satz von F. E. Browder über nichtlineare Wellengleichingen, Math. Z. **141** (1974).

HENRY D.
[1] *Geometric theory of semilinear parabolic equations*, Lecture Notes in Math. #**840**, Springer, New York, 1981.

HERRERO M. A. and VELÁZQUEZ J. J. L.
[1] Blow-up behaviour of one-dimensional semilinear parabolic equations, Ann. Inst. Henri Poincaré, Analyse Non Linéaire **10** (1993), 131–189.
[2] Flat blow-up in one-dimensional semilinear heat equations, Diff. Int. Eq. **5** (1992), 973–997.
[3] Blow-up profiles in one-dimensional, semilinear parabolic problems, Communs. Part. Diff. Eq. **17** (1992), 205–219.
[4] Plane structures in thermal runaway, Israel J. Math. **81** (1993), 321–341.
[5] Generic behavior of one-dimensional blow-up patterns, Ann. Scuola Norm. Sup. Pisa **19** (1992), 381–450.
[6] Approaching an extinction point in one-dimensional semilinear heat equations with strong absorption, J. Math. Anal. Appl. **170** (1992), 353–381.
[7] Some results on blow up for some semilinear parabolic problems, in *Degenerate diffusion (Minneapolis, 1991)*, IMA Vol. Math. Appl. **47**, Springer, New York, 1993, 105–125.

HEYWOOD J. G.
[1] The Navier–Stokes equations: On the existence, regularity and decay of solutions, Indiana Univ. Math. J. **29** (1980), 639–681.

ITO S.
[1] The existence and the uniqueness of regular solution of non-stationary Navier–Stokes equation, J. Fac. Sci. Univ. Tokyo **9** (1961), 103–140.

JOHN F.
[1] Blow-up for quasilinear wave equations in three space dimensions, Communs. Pure Appl. Math. **34** (1981), 29–52.
[2] Formation of singularities in one-dimensional nonlinear wave propagation, Communs. Pure Appl. Math. **27** (1974), 377–405.

[3] Blow-up of solutions of nonlinear wave equations in three space dimension, Manuscripta Math. **28** (1979), 235–268.

JONES C. and KÜPPER T.

[1] On the infinitely many solutions of a semilinear elliptic equation, SIAM J. Math. Anal. **17** (1986), 803–836.

JONES C. K. R. T.

[1] An instability mechanism for radially symmetric standing waves of a nonlinear Schrödinger equation, J. Diff. Eq. **71** (1988), 34–62.

[2] Instability of standing waves for nonlinear Schrödinger equations, Ergod. Th. Dynam. Sys. **8** (1988), 119–138.

JÖRGENS K.

[1] Das Anfangswertproblem in grossen für eine klasse nichtlinearer Wellengleichungen, Math. Z. **77** (1961), 295–308.

KATO T.

[1] *Quasi-linear evolution equations, with applications to partial differential equations*, Lecture Notes in Math. #**448**, Springer, New York, 1975.

[2] On the Cauchy problem for the (generalized) KdV equation, Advances in Math. Supplementary studies, Studies in Applied Mathematics **8** (1983), 93–128.

[3] On nonlinear Schrödinger equations, Ann. Inst. Henri Poincaré, Physique Théorique **46** (1987), 113–129.

KAVIAN O.

[1] A remark on the blowing-up of solutions to the Cauchy problem for nonlinear Schrödinger equations, Trans. Am. Math. Soc. **299** (1987), 193–205.

[2] Remarks on the time behaviour of a nonlinear diffusion equation, Ann. Inst. Henri Poincaré, Analyse Non Linéaire **4** (1987), 423–452.

KELLER C.

[1] Stable and unstable manifolds for the nonlinear wave equation with dissipation, J. Diff. Eq. **50** (1983), 330–347.

KELLER J. B.

[1] On solutions of nonlinear wave equations, Communs. Pure Appl. Math. **10** (1957), 523–530.

KIRANE M. and TRONEL G.

[1] Régularité C^∞ de solutions de problèmes paraboliques semi-linéaires, Publication du Laboratoire d'Analyse Numérique #**84011**, Univ. P. et M. Curie, Paris, 1984.

KLAINERMAN S. and PONCE G.
[1] Global, small amplitude solutions to non-linear evolution equations, Communs. Pure Appl. Math. **36** (1983), 133–141.

LADYZHENSKAYA O. A.
[1] *The mathematical theory of viscous incompressible flow*, Gordon and Breach, 1969.
[2] On the determination of global B-attractors for semi-groups generated by boundary-value problems for nonlinear dissipative partial differential equations, preprint, Steklov Inst., Leningrad, 1987 (in Russian).

LADYZHENSKAYA O. A., SOLONIKOV V. A., and URAL'CEVA N. N.
[1] *Linear and quasilinear equations of parabolic type*, Am. Math. Soc., 1968.

LANGE H.
[1] On some nonlinear Schrödinger equations of Hartree type, Math. Verf. Math. Physik Ba. **21** (1980), 99–111.

LASALLE J. P.
[1] Asymptotic stability criteria, Proc. Symp. Appl. Math. **13** (1962), 299–307.

LERAY J.
[1] Etude de diverses équations intégrales non linéaires et quelques problèmes que pose l'hydrodynamique, J. Math. Pures Appl. **12** (1933), 1–82.
[2] Essai sur les mouvements plans d'un fluide visqueux que limitent des parois, J. Math. Pures Appl. **13** (1934), 331–418.
[3] Sur le mouvement d'un fluide visqueux emplissant l'espace, Acta Math. **63** (1934), 193–248.

LEVINE H. A.
[1] Some nonexistence and instability theorems for formally parabolic equations of the form $Pu_t = -Au + f(u)$, Arch. Rat. Mech. Anal. **51** (1973), 371–386.
[2] Instability and nonexistence of global solutions to nonlinear wave equations of the form $Pu_{tt} = -Au + f(u)$, Trans. Am. Math. Soc. **192** (1974), 1–121.
[3] Nonexistence of global weak solutions to some properly and improperly posed problems of mathematical physics: the method of unbounded Fourier coefficients, Math. Anal. **214** (1975), 205–220.

LIN J. E. and STRAUSS W. A.
[1] Decay and scattering of solutions of a nonlinear Schrödinger equation, J. Funct. Anal. **30** (1978), 245–263.

LIONS P.-L.
[1] Asymptotic behavior of some nonlinear heat equations, Nonlinear Phenomena, Physica **D5** (1982), 293–306.

[2] Structure of the set of steady-state solutions and asymptotic behaviour of semilinear heat equations, J. Diff. Eq. **53** (1984), 362–386.
[3] *Mathematical topics in fluid mechanics, Vol. 1: Incompressible models*, Oxford Science Publications, 1996.
[4] *Mathematical topics in fluid mechanics, Vol. 2: Compressible models*, Oxford Science Publications, 1998.

LOUZAR T.
[1] Comportement à l'infini des solutions radiales d'une équation de la chaleur non linéaire, Thesis, Univ. P. et M. Curie, Paris, 1984.

MARCUS M. and MIZEL V.J.
[1] Every superposition operator mapping one Sobolev space into another is continuous, J. Funct. Anal. **33** (1979), 217–229.

MARSHALL B., STRAUSS W. A., and WAINGER S.
[1] L^p–L^q estimates for the Klein–Gordon equation, J. Math. Pures Appl. **59** (1980), 417–440.

MASUDA K.
[1] On the global existence and asymptotic behaviour of solutions of reaction-diffusion equations, Hokkaido Math. J. **12** (1983), 360–370.

MATANO H.
[1] Asymptotic behavior and stability of solutions of semilinear diffusion equations, Publ. Res. Inst. Math. Sci. (Kyoto Univ.), **15** (1979), 401–454.

MORAWETZ C. and STRAUSS W. A.
[1] Decay and scattering of solutions of a nonlinear relativistic wave equation, Communs. Pure Appl. Math., **25** (1972), 1–31.

MUELLER C. E. and WEISSLER F. B.
[1] Single point blow-up for a general semilinear heat equation, Indiana Univ. Math. J. **35** (1986), 881–913.

NI W. M., SACKS P. E. and TAVANTZIS J.
[1] On the asymptotic behavior of solutions of certain quasilinear parabolic equations, J. Diff. Eq., **54** (1984), 97–120.

PAYNE L. and SATTINGER D. H.
[1] Saddle points and instability of nonlinear hyperbolic equations, Israel J. Math. **22** (1975), 273–303.

PAZY A.
[1] *Semi-groups of linear operators and applications to partial differential equations*, Applied Math. Sciences #44, Springer, New York, 1983.

POLÁČIK P. and RYBAKOWSKI K.
[1] Nonconvergent bounded trajectories in semilinear heat equations, J. Diff. Eq. **124** (1983), 525–537.

PROTTER M. H. and WEINBERGER H. F.
[1] *Maximum principles in differential equations*, Prentice-Hall, 1967.

RABINOWITZ P. H.
[1] Periodic solutions of nonlinear hyperbolic partial differential equations, Communs. Pure Appl. Math. **20** (1967), 145–205.
[2] Free vibrations for a semi-linear wave equation, Communs. Pure. Appl. Math. **31** (1978), 31–68.

REED M. and SIMON B.
[1] *Methods of modern mathematical physics, Vol. II*, Academic Press, 1975.

SCHOCHET H. and WEINSTEIN M. I.
[1] The nonlinear Schrödinger limit of the Zakharov equations governing Langmuir turbulence, Communs. Math. Phys. **106** (1986), 569–580.

SEGAL I. E.
[1] Nonlinear semi-groups, Ann. Math. **2** (1963), 339–364.

SELL G.
[1] Non autonomous differential equations and topological dynamics, Trans. Am. Math. Soc. **127** (1967), 241–283.

SERRE D.
[1] A propos des invariants de diverses équations d'évolution, in *Nonlinear partial differential equations and their applications, College de France Seminar, vol. 3*, H. Brezis and J.-L. Lions (eds), Research Notes in Math. #70, Pitman, 1981, 326–336.

SHATAH J. and STRAUSS W. A.
[1] Instability of nonlinear bound states, Communs. Math. Phys. **100** (1985), 173–190.

SIDERIS T.
[1] Global behaviour of solutions to nonlinear wave equations in three space dimensions, Communs. Part. Diff. Eq. **8** (1983), 1291–1323.
[2] Formation of singularities in solutions to nonlinear hyperbolic equations, Arch. Rat. Mech. Anal. **86** (1984), 369–381.
[3] Nonexistence of global solutions to semilinear wave equations in high dimensions, J. Diff. Eq. **52** (1984), 378–406.
[4] Decay estimates for the three dimensional Klein–Gordon equation and applications, Communs. Part. Diff. Eq. **14** (1989), 1421–1455.

SILI A.
[1] Thèse de 3° Cycle, Université Pierre et Marie Curie, Paris, 1987.

SIMON L.
[1] Asymptotics for a class of nonlinear evolution equation equations, with applications to geometric problems, Ann. Math., **118** (1983), 525–571.

SJÖLIN P.
[1] Regularity of solutions to Schrödinger equations, Duke Math. J. **55** (1987), 699–715.

SMOLLER J.
[1] *Shock waves and reaction diffusion systems*, Springer, New York, 1983.

STRAUSS W. A.
[1] On weak solutions of semilinear hyperbolic equations, An. Acad. Brasil. Ciên. **42** (1970), 645–651.
[2] *Nonlinear invariant wave equations*, Lecture Notes in Physics #**73**, Springer, 1978.
[3] *Nonlinear wave equations*, Regional Conference Series in Mathematics #**73**, Am. Math. Soc., Providence, 1989.

STRAUSS W. A. and VÁZQUEZ L.
[1] Numerical solution of a nonlinear Klein–Gordon equation, J. Comput. Phys. **28** (1978), 271–278.

STRICHARTZ M.
[1] Restrictions of Fourier transforms to quadratic surfaces and decay of solutions of wave equations, Duke Math. J. **44** (1977), 705–714.

TEMAM R.
[1] *Navier–Stokes equations*, North Holland, 1977.
[2] *Infinite dynamical systems in mechanics and physics*, Appl. Math. Sci. #**68**, Springer, New York, 1988.

TSUTSUMI M.
[1] Nonexistence of global solutions to the Cauchy problem for the damped nonlinear Schrödinger equation, S.I.A.M. J. Math. Anal. **15** (1984), 357–366.

TSUTSUMI Y.
[1] Scattering problem for nonlinear Schrödinger equations, Ann. Inst. Henri Poincaré, Physique Théorique **43** (1985), 321–347.
[2] Global solutions of the nonlinear Schrödinger equation in exterior domains, Communs. Part. Diff. Eq. **8** (1984), 1337–1374.
[3] Local energy decay of solutions to the free Schrödinger equation in exterior domains, J. Fac. Sci. Univ. Tokyo, Sect. 1A, Math. **31** (1984), 97–108.

[4] L^2-solutions for nonlinear Schrödinger equations and nonlinear groups, Funk. Ekva. **30** (1987), 115–125.

VEGA L.
[1] Schrödinger equations: pointwise convergence to the initial data, Proc. Am. Math. Soc. **102** (1988), 874–878.

WEINSTEIN M. I.
[1] Nonlinear Schrödinger equations and sharp interpolation estimates, Communs. Math. Phys. **87** (1983), 567–576.
[2] Lyapunov stability of ground states of nonlinear dispersive evolution equations, Communs. Pure Appl. Math. **29** (1986), 51–68.
[3] Modulational stability of ground states of nonlinear Schrödinger equations, SIAM J. Math. Anal. **16** (1985), 567–576.
[4] On the structure and formation of singularities of solutions to nonlinear dispersive evolution equations, Communs. Part. Diff. Eq. **11** (1986), 545–565.
[5] Solitary waves of nonlinear dispersive evolution equations with critical power nonlinearity, J. Diff. Eq. **69** (1987), 192–203.

WEISSLER F. B.
[1] Semilinear evolution equations in Banach spaces, J. Funct. Anal. **32** (1979), 277–296.
[2] Local existence and nonexistence for semilinear parabolic equations in L^p, Indiana Univ. Math. J. **29** (1980), 79–102.
[3] Single point blow-up for a semilinear initial value problem, J. Diff. Eq. **55** (1984), 204–224.
[4] Rapidly decaying solutions of an ordinary differential equation with applications to semilinear elliptic and parabolic partial differential equations, Arch. Rat. Mech. Anal. **91** (1986), 247–266.

YAJIMA K.
[1] Existence of solutions for Schrödinger evolution equations, Communs. Math. Phys. **110** (1987), 415–426.

YAO J.-Q.
[1] Comportement à l'infini des solutions d'une équation de Schrödinger non linéaire dans un domaine extérieur, C. R. Acad. Sci. Paris. **294** (1982), 163–166.

YOSIDA K.
[1] *Functional Analysis*, Springer, New York, 1965.

Index

a priori estimate 58, 135–7
adjoint 22–3
attractor 152

behaviour, asymptotic 120–2, 142
blow-up 72–6, 87–90, 114–21
blow-up alternative 57, 58
Bochner's Theorem 7
boundary regularity 28, 62, 124, 134, 146

compactness 124–41, 143–4, 151
complete metric space 56, 124, 147
conservation of energy 78, 83, 92, 100
continuous, absolutely 13, 14, 15, 53, 55, 139, 141,
contraction
  semigroup 39–41
  strict 56
convexity 6, 148

damping 50–55, 134–41
dependence, continuous 59, 100
differentiable, differentiability 38, 51, 60, 62, 78
differential inequality 72, 73, 75, 87, 88, 116, 117, 120, 125, 131–3, 137, 139
distribution 2
  vector-valued 10
domain
  bounded 27, 43, 46, 62, 68, 74, 124, 134, 141, 146

of an operator 18
dynamical system 142, 143, 145

eigenfunction 72
embedding, compact 3, 151
energy, *see* conservation of energy
equation
  inhomogeneous 50–55
  non-autonomous 50–55, 134–41
  parabolic, *see* heat equation
  with second member, *see* equation, non-autonomous
equilibrium point 144, 148, 152
estimate, uniform 67, 68, 84, 112–13, 130, 136, 139
existence
  global 58, 65–71, 76, 83–6, 112––14
  local 56–9, 64–5, 76, 82–3, 100–12
extrapolation 163

forcing 50–55, 134–41
function
  integrable 7–8
  measurable 4–6
functional analysis 1

Gagliardo–Nirenberg's inequality 3, 112
generator 39
graph, closed 18, 20
Gronwall's lemma 55, 125

growth condition 65, 70, 84, 86, 112, 119, 127, 130, 133, 138, 149

heat equation 42–7, 62–77, 124–30, 134–7, 146–9, 158–63, 164–8
Hilbert complex space 25–6

Hille–Yosida–Phillips Theorem 40

invariance principle 143–4
isometry group 37, 41, 47–9, 61, 78, 91, 137, 149

Klein–Gordon equation, *see* wave equation

Lax–Milgram Theorem 1, 26–7
Liapunov function 143–4, 147, 150
  strict 144, 147, 150
Lipschitz condition, local 55, 57, 62, 63, 79–81, 100, 145–6, 149

maximum principle, 65–7

ω-limit set 142, 143, 145
operator
  closed 18, 20
  dissipative 19
  m-dissipative 18–32, 33–5, 38–41
  self-adjoint 24, 26, 32, 35
  skew-adjoint 24, 26, 29–32, 37, 61, 78

Pettis' Theorem 5
point
  fixed 57
  hyperbolic 156

regularity 28, 33, 35–7, 39, 41–3, 47, 48, 60–1, 62, 78, 100

Schrödinger equation 47–9, 91–123
set, connected 143
smoothing effect 37
Sobolev embeddings 3
Sobolev space
  2, 13–17
solution
  bounded, *see* estimate, uniform
  global, *see* existence, global
  maximal 57
  stationary 157
stability 154
  asymptotic 154
  exponential 158–64
Strichartz' estimates 96–100

trajectory 142
  bounded, 144–7
  relatively compact 143

uniqueness 56, 106

variation of the parameter formula 50

wave equation 47, 78–90, 130–3, 137––41, 149–52, 163–4